COMBINATORIAL GEOMETRY IN THE PLANE

Hugo Hadwiger

and

Hans Debrunner

Translated by Victor Klee

DOVER PUBLICATIONS, INC.

Mineola, New York

Copyright

Bibliographical Note

This Dover edition, first published in 2015, is an unabridged republication of the work, translated by Victor Klee, and originally published in the "Athena Series: Selected Topics in Mathematics" by Holt, Rinehart and Winston, Inc., New York, in 1964. The translator has also supplied a new Chapter 11 ("Further Development of Combinatorial Geometry") and other additional material. The original German language edition was first published as *Kombinatorische Geometrie in der Ebene,* Monographies de L'Enseignement Mathematique, No. 2, Geneva.

This edition is published by special arrangement with Cengage Learning, Inc., Belmont, California.

Library of Congress Cataloging-in-Publication Data

Hadwiger, Hugo.
[Kombinatorische Geometrie in der Ebene. English]
Combinatorial geometry in the plane / Hugo Hadwiger and Hans Debrunner. — Dover edition.
pages cm
Original translated by Victor Klee.
Originally published: New York : Holt, Rinehart and Winston, 1964.
Includes bibliographical references and index.
ISBN-13: 978-0-486-78996-5
ISBN-10: 0-486-78996-9
1. Convex domains. 2. Combinatorial geometry. I. Debrunner, Hans. II. Title.

QA603.H313 2015
516'.13—dc23

2014025108

Manufactured in the United States by Courier Corporation
78996901 2015
www.doverpublications.com

Preface

On both high school and collegiate levels, the content and approach of our mathematics curriculum have changed materially in recent years. Though many pedagogic improvements have resulted, three important problems are unsettled. One of these, widely recognized by the professional committees that have sparked the curriculum change, is "What should be done about geometry?" Powerfully influenced by Euclid's example of over two thousand years ago, some mathematicians feel that geometry is still the ideal subject for introducing the student to rigorous thought. Others argue that the logical structure of geometry is much too complicated for this purpose, that Euclid had no choice because geometry was *the* mathematics of his day, and that in any case the traditional axiomatic approach to geometry is inefficient.

The other two questions are "How should we encourage the student to develop his mathematical intuition, and how help him to understand the nature of mathematical research?" It seems to us that these two questions are closely related and have a common answer. We should expose the student to some material that has strong intuitive appeal, is currently of research interest to professional mathematicians, and in which the student himself may discover interesting problems that even the experts are unable to solve. There are two areas in mathematics that clearly conform to this description. One is elementary number theory and the other is the sort of geometry that forms the content of this book, largely associated with the notion of convexity. We submit the present book as a tentative answer to all three of the questions mentioned above. Evidence favoring this sort of solution is provided by the use in Russian schools of several works on convex bodies, especially the book by Yaglom and Boltyanskiĭ [114] which is an excellent companion volume for the present one.*

In all essentials, the translation adheres closely to the original text. At the authors' invitation, the translator has added a number of references and also an eleventh chapter that surveys recent developments in the field. This is based partly on his joint paper with Ludwig Danzer and Branko Grünbaum, whose collaboration was also useful to the authors in preparing the original text.

Seattle, Washington
October, 1963

V.K.

*The book by Yaglom and Boltyanskiĭ has been translated into both German and English, the latter edition being published by Holt, Rinehart and Winston, Inc. The German edition of the present book was based on the article "Ausgewählte Einzelprobleme der kombinatorischen Geometrie in der Ebene" by H. Hadwiger in L'Enseignement Mathematique (2), **1**, 56–89 (1955), translated into French by J. Chatelet, *ibid.* **3**, 35–70 (1957).

Contents

COMBINATORIAL GEOMETRY IN THE PLANE

Introduction

There are various mathematical subjects in which elementary exercises lead at once to more advanced and partially unsolved problems, so that the simplest matters of school mathematics are closely tied to those that are of scientific interest and are studied by specialists. A key point in this is that the two professional levels are not, as is usual, separated from each other by highly developed advanced theories and stratified scales of ideas.

Such a subject is combinatorial geometry, which has an especially simple character when restricted to the plane. Its problems are directly connected with the basic ideas of elementary plane geometry and are based on various primitive relations and processes, such as those of inclusion, intersection, decomposition, and so forth, and on the combinatorial possibilities associated with these relations and processes.

The subject is related to combinatorial topology; however, genuine topological considerations remain well in the background, and the difficulties are those of elementary geometry. As was more fully explained by H. Hopf [47], there is a certain reciprocal relationship between the metric and topological viewpoints in combinatorial geometry.

The gathering of numerous special problems that we have undertaken is not, however, totally restricted to the methods of combinatorial geometry; these form only a tiny nucleus of a complex of questions that has exerted a special attraction because of the simple and basic nature of its subject matter, and the purely combinatorial aspect of the necessary inferences.

How a person familiar only with elementary concepts can pose problems so as to develop this taste, and accustom himself to a change that leads from the methods and topics of the familiar classical domain to those of a more currently oriented research area with exciting new possibilities — this will be brought home to the reader by the examples collected here.

Except for basic material from elementary geometry and the theory of real numbers, little is needed in the way of previous knowledge; a certain familiarity with set-theoretical reasoning is useful, and the notion of a plane point set is important. Most of the notation will be briefly explained.

In Part I, selected theorems are assembled, arranged into groups of related propositions without proof but with rather detailed comments and with references to the literature. The proofs, often only briefly indicated, follow in Part II. Thus the reader will have the opportunity to practice the search for and execution of his own ideas of proof. Through the numerous references, especially interested readers may also find their way to the current research literature and may even pursue the unsolved problems that are suggested.

With these selected special problems we hope to stimulate an intensive study of the fascinating questions of combinatorial geometry and to bring into active being the close contact that exists between school mathematics and scientific research in this domain.

Part I

1. Incidence of Points, Lines, and Circles

The propositions of this first small group deal with incidence relations among points, lines, and circles, and thus pertain to combinatorial elementary geometry.

1. If a finite set of points is such that on the line determined by any two of the points there is always a third point of the set, then all the points lie on a single line.

This theorem was conjectured in 1893 by J. J. Sylvester [102]. A short proof due to T. Gallai (Grünwald) is given by N. G. de Bruijn–P. Erdös [9], where the result appears as a corollary of a purely combinatorial theorem. For further proofs, generalizations, and variants see P. Erdös [16], H. S. M. Coxeter [10], G. A. Dirac [13], and Th. Motzkin [76].

2. If a finite set of lines is such that through the intersection point of any two of the lines there always passes a third line of the set, then all the lines pass through a single point.

Propositions **1** and **2** are no longer true if the sets of points or lines are infinite. This is demonstrated for both theorems simultaneously by the example of the regular countably infinite system of points and lines pictured in Figure 1.

3. If a finite set of points, not all collinear, is such that on the circle determined by any three of the points there is always a fourth point of the set, then all the points lie on a single circle.

A set is said to be *bounded* if it lies in some circular disk of finite radius, *closed* if it includes all its *limit points*, where a point p is a limit point of a set X if every disk centered at p includes a point of X different from p. In both hypothesis and conclusion, the following theorem is closely related to Proposition **3**.

4. If a bounded closed set of points is such that the axis of symmetry for any two of its points is always an axis of symmetry for the entire set, then all the points lie on a single circle.

It is easy to see that Propositions **3** and **4** are invalid for sets that are both infinite and unbounded. Indeed, it suffices to consider the entire plane as the

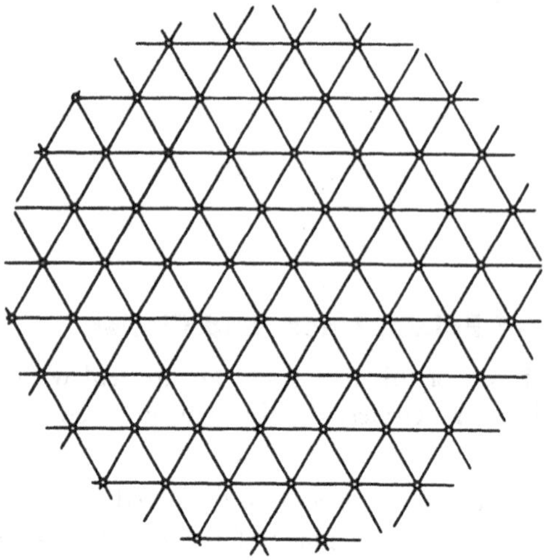

Fig. 1

set of points in question. An example consisting of countably many points may be constructed as follows: Let A_0 be a set of four points not lying on any line or circle. Then let an increasing sequence of finite sets $A_n (n = 0,1, \cdots)$ be constructed recursively by setting

$$A_n = \phi(A_{n-1}) \qquad (n = 1,2, \cdots),$$

where $\phi(A)$ denotes the union of all reflections of A in the various lines that are axes of symmetry for pairs of points in A. As one easily verifies, the union S of all the sets A_n is a countably infinite set with the desired property of symmetry. On the circle determined by any three points of S, there is always a fourth point of S as long as the three points do not determine an equilateral triangle (and even when they do if the construction ϕ is slightly extended).

2. Integral Distances. Commensurable Angles

We present next a group of propositions in which a role is played by the integral or rational nature of certain distances.

The set of all points, whose coordinates with respect to an orthogonal coordinate system are both integers, forms the plane *unit lattice;* its points are called *lattice points.*

5. If n lattice points form a regular n-gon $(n > 2)$, then $n = 4$; thus the square is the only regular polygon that can be embedded in the unit lattice.

An ingenious proof of this was given by W. Scherrer [93]; for the case $n = 3$, see also G. Polya–G. Szegö [82], **2**, page 156, problem 238.

As is seen in Figure 2, a square can be embedded in the lattice in ways other than the obvious ones.

For the angles of embedded rhombi the following is known.

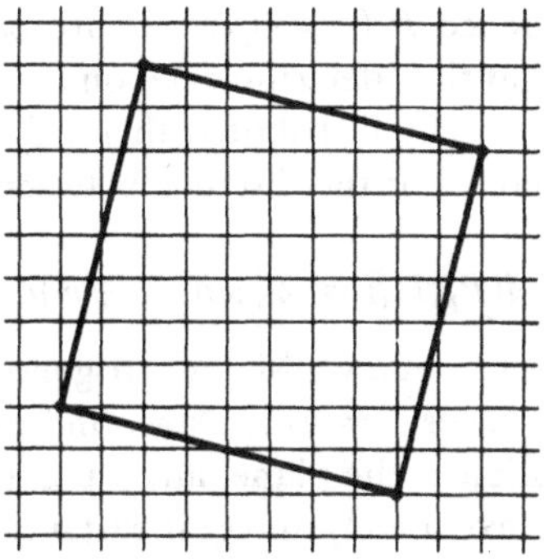

Fig. 2

6. If four lattice points determine a nonsquare rhombus with angle α, the quotient α/π is irrational; thus the square is the only rhombus that has angles commensurate with right angles and that can be embedded in the unit lattice.

Closely connected with this is an assertion about angles in Pythagorean triangles, that is, in right triangles whose sides have integral lengths.

7. If α is an acute angle of a Pythagorean triangle, the number α/π is irrational.

Propositions **6** and **7** are geometric corollaries of the following trigonometric theorem (Hadwiger [33]).

8. If $0 < \alpha < \pi/2$ and the number $\cos \alpha$ is rational, then either $\alpha = \pi/3$ or α/π is irrational.

9. If an infinite set of points is such that all of its points are at integral distances from each other, then all of the points lie on a single line.

This theorem of P. Erdös [17] may serve as prototype for a certain category of results that are especially appealing in that strong and unexpected consequences are drawn from the simplest assumptions.

Remarkably, Proposition **9** does not imply the existence of a number k_0 such that the conclusion always holds when the number k of points with exclusively integral distances is greater than k_0. In fact, for each k there are such sets that are not linear, and even ones that have no three collinear points. Such point sets have been constructed repeatedly by M. Altwegg [1], A. Müller [77], F. Steiger [98], and others.

Following an idea of A. Müller, it is possible to describe a countably infinite point set that is dense in the unit circle, that is, intersects every arc of the circle, and has the property that all of its points are at rational distances from each other. Let P_n be the point with polar coordinates $\rho = 1, \theta = 2n\phi$, where ϕ is such that $\cos \phi = 4/5$, so that ϕ/π is irrational by Proposition **8**.

The points of the sequence $P_n(n = 0,1, \cdots)$ are pairwise distinct, and the countably infinite set that they determine lies on the unit circle. It is dense in the circle and even uniformly distributed there, though we shall not define this notion precisely here. For the distance between any two of these points we have

$$d(P_n,P_m) = 2 \mid \sin (n - m)\phi \mid,$$

whose rationality follows from a well-known trigonometric identity and from the fact that $\sin \phi = 3/5$, $\cos \phi = 4/5$. Now, under an appropriate dilation, a similarity transformation of the plane, any k points of this set give rise to a set of k points having exclusively integral distances. Still no three of the points are collinear.

3. Hull Formation. Separability

The following group of propositions is concerned with hull formation and separation for plane point sets. First, some definitions. A set is called *convex* if with any two of its points it contains the entire line segment joining them. The *convex hull* of a set is the smallest convex set containing it. Alternatively, the convex hull is the intersection of all convex sets that contain the given set. A point is *interior* to a set if the set contains some circular disk centered at the point.

10. A point is in the convex hull of a set if, and only if, it is already in the convex hull of three or fewer points of the set.

From this it follows that the convex hull is identical with the union of all triangles whose vertices belong to the given set, or with a similar union of segments when the set is contained in a line.

11. A point is an interior point of the convex hull of a set if, and only if, it is interior to the convex hull of four or fewer points of the set.

Propositions **10** and **11** are the two-dimensional special cases of useful theorems of E. Steinitz [99] and W. Gustin [30]. See also O. Hanner–H. Rådström [41] and C. V. Robinson [90].

Two sets will be called *separable* if there is a line that does not intersect either set and that separates them from each other; the two sets then lie interior to the two half planes that are determined by the line. The following criterion for separability is due to P. Kirchberger [55] (see also H. Rademacher–I. J. Schoenberg [83]).

12. Two bounded closed sets are separable if, and only if, any two of their subsets whose union includes at most four points are separable.

13. Each set that includes at least four points can be decomposed into two nonempty disjoint subsets that are not separable.

In this connection, see F. W. Levi [69] and R. Rado [86].

4. Helly's Theorem. Transversal Problems for Ovals

We turn now to a circle of questions centered around the famous theorem of Helly. The numerous variants, Helly type theorems, which as a rule refer to ovals, form a very typical part of the combinatorial geometry of convex sets.

By an *oval* we understand here a bounded, closed convex set.

14. If each three ovals of a finite or infinite family of ovals have a common point, then all the ovals of the family have a common point.[1]

This is the plane case of the well-known theorem of Helly. See E. Helly [42], J. Radon [88], D. König [64], and others. As one sees immediately from the simplest examples, the number three cannot be replaced by two. This is possible, however, with a strong assumption on the shape of the ovals. Thus the following variant is known.

15. If each two rectangles of a family of "parallel" rectangles, that is, with sides parallel to the coordinate axes, have a common point, then all the rectangles of the family have a common point.

In contrast, an oval that is not a parallelogram can be placed in three positions so that every two of the translates have a common point, but not all three. For parallelograms this is not possible. Thus parallelograms are characterized by a slightly modified version of the property expressed in Proposition **15.** See B. Sz.–Nagy [78] and especially Proposition **87** below.

A corollary of Proposition **15** is Helly's theorem for the line.

16. If each two segments of a family of segments (in the line) have a common point, then all the segments of the family have a common point.

Theorems of Helly type for the circle are closely related to Proposition **16** and are useful for many applications; instead of ovals, these involve circular arcs, which obviously should lie on the same circle.

17. If a family of circular arcs, all smaller than a semicircle, is such that each three of the arcs have a common point, then all the arcs of the family have a common point.

The condition on the size of the arcs cannot be weakened, since the proposition is already false for semicircles. In fact, if four semicircles arise from two different pairs of antipodal points of the circle, then each three have a common point but not all four. Also, the number three in Proposition

[1]When discussing a family of sets or points, we often have occasion to require that for each way of choosing k (not necessarily distinct) members from the family, the k members chosen have, collectively, a certain property (such as a nonempty intersection). We express this succinctly by saying that each k members have the property.

17 cannot be replaced by two; of three equal arcs that just cover the entire circle, each two have a common point but not all three. On the other hand, the following is known.

18. If a family of circular arcs, all smaller than one third of a circle, is such that each two of the arcs have a common point, then all the arcs of the family have a common point.

For the next result, we drop all assumptions on the size of the arcs.

19. If a family of circular arcs is such that each two of the arcs have a common point, then there is an antipodal pair of points such that each arc of the family includes at least one point of the pair.

In other words, there is a diameter of the circle that intersects all the arcs. Theorems of this sort were found by C. V. Robinson [90] and A. Horn–F. A. Valentine [50], among others. Some beautiful applications, such as those we shall describe later, were discovered by P. Vincensini [111].

20. If an oval can always be translated so that it is contained in the intersection of any three members of a family of ovals, then it can also be translated so that it is contained in the intersection of all the ovals of the family.

21. If an oval can always be translated so that it intersects any three members of a family of ovals, then it can also be translated so that it intersects all the members of the family.

22. If an oval can always be translated so that it contains any three members of a family of ovals, then it can also be translated so that it contains all of the members.

These are the plane cases of more general (higher-dimensional) variants of Helly's theorem, formulated by P. Vincensini [110] and V. L. Klee, Jr. [58]. For the validity of these results it is essential that the ovals should only be translated and not rotated. If the translation group is replaced by the group of rigid motions, then all three assertions become false.

We illustrate this more fully by an example connected with Proposition **21.** Consider the family of n circular disks ($n > 2$) whose centers have polar coordinates $\rho = 1$ and $\theta = 2k\pi/n$ ($k = 1, \cdots, n$), and whose radii are all equal to $\cos^2(\pi/n)$ or $\cos^2(\pi/n) + \cos^2(\pi/2n) - 1$ according to whether n is even or odd. As can be verified, a segment (degenerate oval) of length 2 can always be placed so as to intersect any $n - 1$ disks of the family but not to intersect all n of them. For this the segment must be suitably translated *and* rotated. The case $n = 8$ is illustrated by Figure 3.

23. If a family of ovals is such that each two of its members have a common point, then through each point of the plane there is a line that intersects all the ovals of the family.

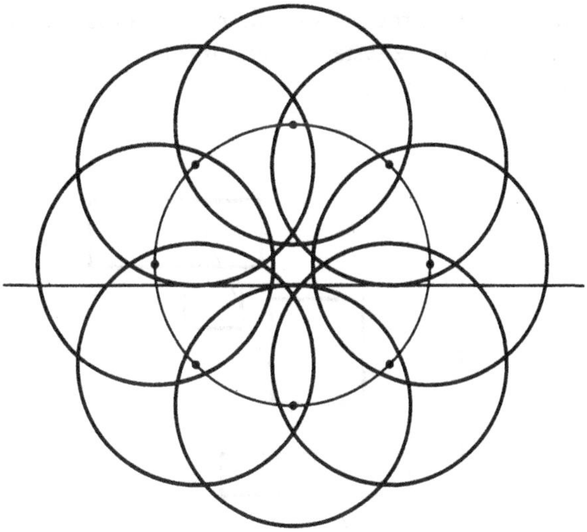

Fig. 3

24. If a family of ovals is such that each two of its members have a common point, then for each line in the plane there is a parallel line that intersects all the ovals of the family.

Propositions **23** and **24** are the plane cases of more general theorems of A. Horn [49] and V. L. Klee, Jr. [56]; they answer the question as to what can replace the conclusion of Helly's theorem when the number three is replaced by two in its hypothesis.

One may ask whether points can be replaced by lines in Helly's theorem, in the sense that an assertion of the following form is correct: If each h members of a family of ovals are intersected by a line, then there is a line that intersects all the ovals of the family. Does there exist such a Helly "stabbing number" h?

The answer is negative! L. A. Santaló [91] has already observed that for each natural number $n > 2$ it is possible to construct a family of n ovals so that each $n - 1$ members of the family admit a common transversal, but not all n. This is illustrated also by our example in connection with Proposition **21**. Theorems of the sort mentioned can be established only under supplementary conditions on the shape and position of the ovals involved. For example, Santaló proved that all of the rectangles of a family of parallel rectangles are intersected by a line if this is the case for each six rectangles of the family. We add here the following assertion, which contains Santaló's theorem.

25. If each three rectangles of a family of parallel rectangles are intersected by an ascending line, then there is an ascending line that intersects all the rectangles of the family.

We assume here that the sides of the rectangles are parallel to the axes of an orthogonal coordinate system; a line is said to be ascending if its slope is nonnegative. See Figure 4.

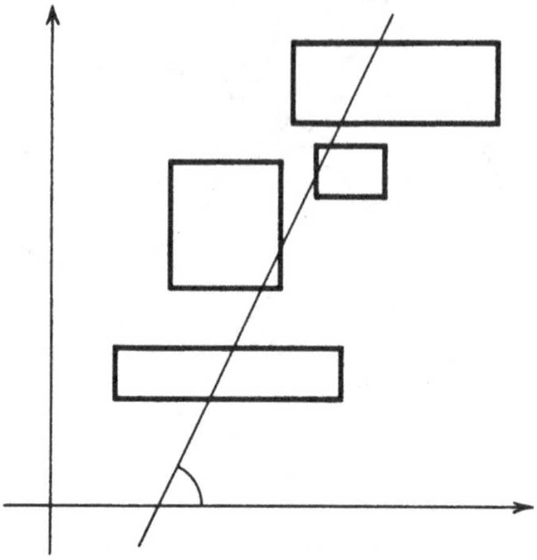

Fig. 4

The earlier example (Figure 3), showing the nonexistence of a Helly stabbing number in the most general case, has the conspicuous feature that the ovals (disks) partially cover each other. Thus it is natural to ask whether a Helly stabbing number can be specified under the assumption that the

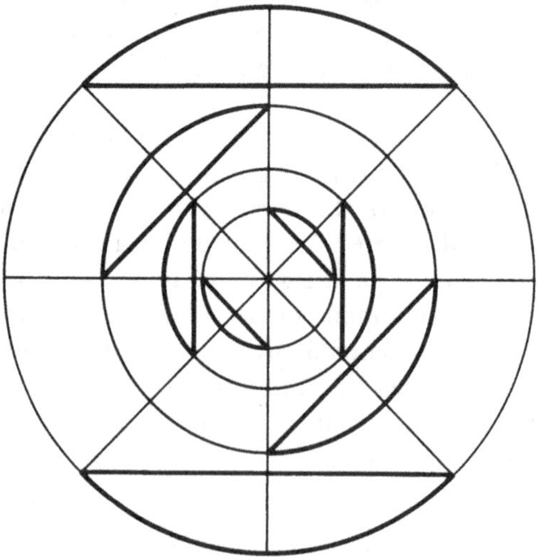

Fig. 5

ovals are pairwise disjoint. The answer to this question, posed by V. L. Klee, Jr. [59], is again negative.

We construct an example, a rosette of circular segments, in order to establish this claim. Suppose $n > 1$. Let S_i and $S_i^*(i = 1, \cdots, 2n)$ be a family of $4n$ circular segments from the $2n$ concentric circles $K_i(i = 1, \cdots, 2n)$ with center Z and radii $R_i(i = 1, \cdots, 2n)$, where S_i and S_i^* are symmetrically located with respect to Z. For the radii we assume initially only that $0 < R_i \leqq R_{i+1}$. The segments of the circle K_i are determined by the polar coordinates of the points of their circular arcs:

$$S_i\colon \quad \rho = R_i;\ (i - n + 1)\left(\frac{\pi}{2n}\right) \leqq \theta \leqq (i + n - 1)\left(\frac{\pi}{2n}\right)$$

$$S_i^*\colon \quad \rho = R_i;\ (i + n + 1)\left(\frac{\pi}{2n}\right) \leqq \theta \leqq (i + 3n - 1)\left(\frac{\pi}{2n}\right).$$

With a view to future applications, we wish to establish some properties of our rosette:

Property 1. The radii R_i can be chosen so that the $4n$ segments are pairwise disjoint; this requires only that the R_i's increase rapidly enough. Figure 5 shows a rosette of this sort for $n = 2$.

Property 2. There is no line that intersects all $4n$ segments. Consider first a line through Z. Because of the $4n$-fold rotational symmetry in the angular coordinate, we may assume that the line's angle of inclination θ is such that $0 \leqq \theta < \pi/2n$. But the segments S_n and S_n^* are not intersected by such a diametral line, and any line parallel to it must miss S_n or S_n^*.

Property 3. There is no point that belongs to all $4n$ segments. This is a trivial consequence of Property 2.

Property 4. In the case $R_i = R(i = 1, \cdots, 2n)$, each $2n - 1$ of the pairs of segments have a common pair of antipodal points. It suffices to consider all pairs other than S_n and S_n^*. The two points $\rho = R$, $\theta = 0$ and $\rho = R$, $\theta = \pi$ belong to them.

Property 5. In the case $R_i = R(i = 1, \cdots, 2n)$, there is no pair of antipodal points belonging to all pairs of segments. This is a trivial consequence of Property 2.

Property 6. Each $2n - 1$ segments are intersected by a line through Z. This is a corollary of Property 4, but here the equality of the radii is irrelevant, so the assertion holds also when the segments are pairwise disjoint.

Property 7. In case $R_i = R(i = 1, \cdots, 2n)$, for each selection of $2n - 1$ segments there are two points such that each of the selected segments includes at least one of the two points. This is a corollary of Property 4.

Property 8. There do not exist two points such that each of the $4n$ segments includes at least one of the two points. This is a corollary of Property 2.

From Properties 1, 2, and 6 there results the negative answer to the question discussed above. The next two propositions, and especially the material of Chapter **10** below, show to what extent the existence of a common transversal can be deduced from Helly type assumptions in the presence of various supplementary conditions. And the same rosette will make it possible to demonstrate the nonexistence of other theorems of Helly type that will occasionally be considered.

In connection with a paper by L. A. Santaló [92], Th. Motzkin communicated a counterexample to the following statement: If each h members of a family of pairs of ovals have a common point, then there is a point common to all the pairs of ovals. Our rosette also refutes this; it is shown in the case of equal radii by Properties 4 and 5.

V. L. Klee, Jr. [57] once asked about the existence of a Helly "piercing number" h for which the following statement is true: If for each h members of a family of ovals there are two points such that each of the h includes at least one of the points, then the same holds for the entire family of ovals. Again there exists no such theorem; our rosettes show this; it follows from Properties 7 and 8 in the case of equal radii.

On the other hand, there is the following result on ovals that are similar and similarly situated.

26. If each four members of a family of homothetic ovals admit a common transversal, then there are four lines, parallel or orthogonal in pairs, such that each of the ovals is intersected by at least one of the lines.

We will end the present group of propositions of Helly type with a variant discovered by P. Vincensini [111]. A system of ovals will be called *totally separable* if there exists a direction such that each line in this direction intersects at most one oval of the system. Then pairwise disjoint parallel strips can be formed in the plane in such a way that each strip contains exactly one oval from the system. See Figure 6.

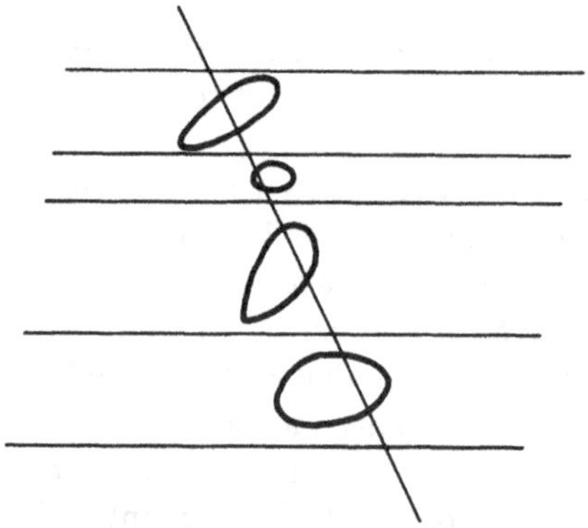

Fig. 6

27. If each three members of a totally separable system of ovals admit a common transversal, then there is a transversal common to all of the ovals of the system.

The stabbing number given by P. Vincensini was $h = 4$. V. L. Klee, Jr. [60] noticed that the theorem could be sharpened by reducing the number to $h = 3$.

A corollary of Proposition **27** is the theorem of L. A. Santaló [91] (see also H. Rademacher–I. J. Schoenberg [83]), according to which there is a transversal common to all the members of a family of parallel segments if the same is true for each three segments from the family.

In view of Proposition **27** it is interesting to inquire what further properties of a system of ovals lead to its total separability. In this connection we mention the condition that the ovals satisfy a transversality condition and in addition are sufficiently sparsely distributed in the plane; this can be described in terms of the size of the viewing angle. See Figure 7.

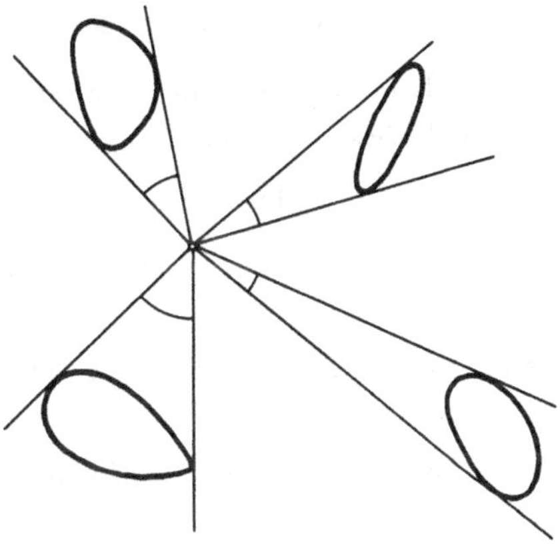

Fig. 7

28. If the ovals of a system are so sparsely distributed that from each point in the plane at most one of the ovals subtends an angle of $\pi/3$ or more, and if each three ovals of the system admit a common transversal, then the system is totally separable, and there is a transversal common to all of the ovals of the system.

5. Covering Problems

We come now to an extensive group of propositions in which the behavior of coverings and especially the possibility of covering by special figures are considered. Also included are some related questions on intersection possibilities.

For a first type of such theorems, the well-known theorem of H. W. E. Jung [51] on the size of the circumcircle of a set of given diameter is typical. We shall begin with some explanations.

Recall that a point set is called *bounded* if it can be covered by a disk. In connection with the discussion below, a set of lines will be called *bounded* if it includes no parallel lines and the set of all intersection points of pairs of lines from the set is bounded.

The *circumradius* of a bounded point set is the radius of the smallest closed disk that includes all points of the set. Similarly, we define the *intersection radius* of a bounded set of lines to be the radius of a smallest closed disk that intersects all lines of the set.

The *diameter* of a bounded point set is the least upper bound of the set of distances realized between pairs of points of the set. Correspondingly, we define the *diameter* of a set of lines as the diameter of the set of all intersection points of the various pairs of lines involved.

29. If each three points of a bounded point set can be covered by a disk of radius R, then some such disk covers the entire set.

30. If each three lines of a bounded set of lines are intersected by some disk of radius R, then some such disk intersects all lines of the set.

These results are special cases of Proposition **21**.

31. For the circumradius R of a point set of diameter $D = 1$, we have $R \leqq 1/\sqrt{3}$.

This is the plane case of Jung's theorem. See the detailed account of H. Rademacher–O. Toeplitz [84].

32. For the intersection radius r of a set of lines of diameter $D = 1$, we have $r \leqq 1/2\sqrt{3}$.

This result is dual to Jung's theorem.

33. A point set of diameter $D = 1$ can be covered by an equilateral triangle of side $s = \sqrt{3}$.

34. A point set of diameter $D = 1$ can be covered by a regular hexagon of side $s = 1/\sqrt{3}$.

A set that has the property that each point set of diameter $D = 1$ can be covered by it is called a *universal cover*. In this sense the disk of radius $R = 1/\sqrt{3}$ is a universal cover. By Propositions **33** and **34** the regular n-gon circumscribed about a disk of diameter $D = 1$ is a universal cover for $n = 3$ or $n = 6$. Proposition **33** is the plane case of a counterpart to Jung's theorem, proved for arbitrary dimension by D. Gale [24]. Proposition **33** is due to J. Pál [81].

35. Every point set of diameter $D = 1$ can be covered by three point sets each of diameter $\sqrt{3}/2$ or less.

This is a sharpening, given by D. Gale [24], of K. Borsuk's theorem [8] to the effect that a plane point set can always be decomposed into three parts each of smaller diameter than the original set. Proposition **52** implies that a finite plane point set can be decomposed into three parts of smaller diameter.

The question considered by Borsuk deals more generally with bounded sets in k-dimensional space, and the conjecture, as yet unproved for $k > 3$, is that such a set can always be decomposed into $k + 1$ sets of smaller diameter. For $k = 3$ the Borsuk theorem was first proved by H. G. Eggleston [14, 15]. Shorter proofs, depending on the decomposition of a suitable universal cover, were found by B. Grünbaum [27] and A. Heppes [43]. A. Heppes and P. Revesz [45] gave a special proof for convex polyhedra, using Euler's theorem on polyhedra.

Under additional hypotheses, Proposition **35** can be sharpened in one respect. First some explanations. A *support line* of an oval is a line that intersects the oval, but such that the oval lies in one of the closed half planes determined by the line. Through each boundary point of an oval there passes at least one support line. If a boundary point admits only one support line, it is called *regular*. Further, by a *strip* of breadth b we mean a closed part of the plane consisting of all points that lie between two parallel lines at distance b from each other.

36. An oval of diameter $D = 1$, whose boundary points are all regular and that can be covered by a strip of breadth $b < 1$, can be covered by two point sets each of diameter <1.

By considering an equilateral triangle of unit side and a circular disk of unit diameter, it is shown that neither of the supplementary hypotheses can be abandoned.

H. Lenz [67, 68] has obtained various results on the magnitudes of diameters for decompositions of point sets into a prescribed number of parts.

For covering by strips, S. Straszewicz [100] showed the following.

37. If an oval can be covered by two strips of breadths a and b, it can be covered by a single strip of breadth $a + b$.

This is a simple case of a 1932 conjecture of A. Tarski, according to which an oval that can be covered by n strips of breadths $a_1, \cdots, a_n$, can also be covered by a single strip of breadth $a_1 + \cdots + a_n$. The correctness of this well-known conjecture, known as the "plank problem," was established by T. Bang [4, 5]; additional proofs were found by W. Fenchel [23] and D. Ohmann [80]. In this connection, the following theorem of W. Blaschke [7] may also be mentioned.

38. If an oval cannot be covered by any strip of breadth 1, it contains a circle of radius $r = 1/3$.

In this connection, see also I. M. Yaglom–V. G. Boltyanskiĭ [114], problem 17.

The more the various ovals of a system are drawn together, the more completely they can be covered by a single given oval. This decreases the possibility that the members of the system can be separated by a line. We say here, in connection with the notion of separability defined just before Proposition **12**, that a system of ovals is *separable* if there is a line that intersects none of the ovals, but such that each of the two open half spaces determined by the line contains an oval of the system. Figure 8 shows a separable system of six ovals together with the convex hull.

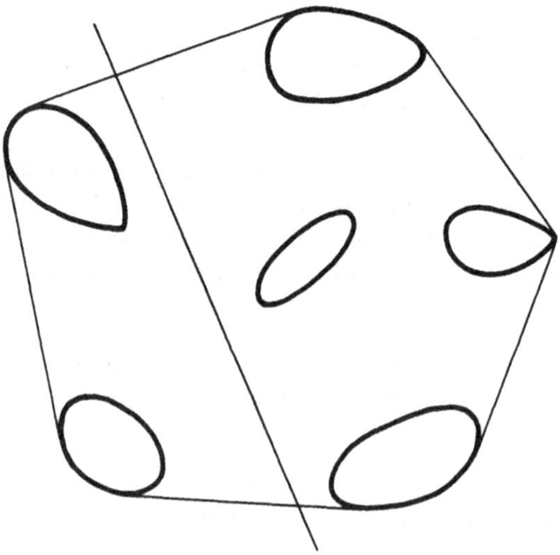

Fig. 8

The ovals of a nonseparable system must obviously lie rather close together; in fact, they can all be covered by an oval whose measurements are comparable with theirs.

39. A nonseparable system of n ovals of circumferences $L_1, \cdots, L_n$ can be covered by an oval of circumference $L_1 + \cdots + L_n$.

This and similar results are found in H. Hadwiger [35]. An interesting special result for disks, not obtainable by the same simple reasoning, is due to A. W. Goodman–R. E. Goodman [25]. It asserts that a nonseparable system of n disks of radii $R_1, \cdots, R_n$ can be covered by a single disk of radius $R_1 + \cdots + R_n$.

The next propositions deal with covering by the interiors of point sets. For this we recall that an oval is, by definition, closed; if it has interior points it is said to be *proper*, otherwise to be *degenerate*. The *interior* of a point set is the set of all interior points of the set. However, in the next two lemmas concerning circular arcs, by the *interior of an arc* we shall mean the set of all of its points except the end points. We say that two sets *do not overlap* if their interiors are disjoint. An example of nonoverlapping sets is given by the lattice arrangement of hexagons in Figure 9, and also by the ovals inscribed in them.

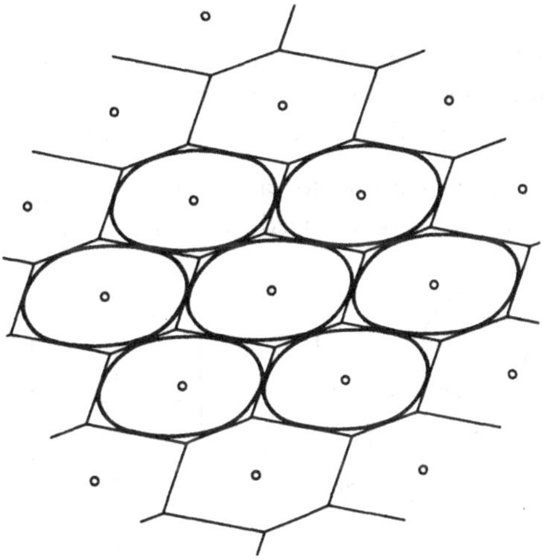

Fig. 9

40. If M is a family of closed nonoverlapping arcs of a circle, each of positive length but smaller than a semicircle, then there exist three semicircles such that each arc of M lies in at least one of the three semicircles, and the three "mid-points" (poles) of these semicircles do not all lie in any closed semicircle. More exactly, this assertion is false if M consists of four arcs that are antipodal in pairs and cover the entire circle; otherwise it is true.

For this proposition there is a polar one, as follows.

41. If M is a family of closed arcs of a circle, each of positive length but smaller than a semicircle, and if the union of any two arcs in M always contains an antipodal pair of points, then there are three points of the circle, not contained in any closed semicircle, such that each arc of M has at least one of these points in its interior. More exactly, this assertion is false if M consists of four nonoverlapping arcs that are antipodal in pairs and cover the entire circle; otherwise it is true.

If one attempts to cover an oval with translates of itself, where only the interior may be used for the covering, it turns out that the parallelogram's behavior is worse than that of any other oval. This result of F. W. Levi [71] can be expressed more exactly as follows.

42. A proper oval A can be covered by three of its translates $A_i (i = 1,2,3)$ in such a way that A lies entirely in the interior of the union of the A_i's. More exactly, this assertion is false if A is a parallelogram; otherwise it is true.

Dually, one may try to intersect an oval with nonoverlapping translates of itself.

43. Let A be a proper oval and let $n(A)$ denote the maximum number of nonoverlapping translates of A that can be arranged so as to intersect A. Then $7 \leqq n(A) \leqq 9$, where equality on each side is realized by a proper oval.

In fact, it is easy to see that for a disk K and a square Q, the extremes $n(K) = 7$ and $n(Q) = 9$ are realized; this is illustrated in Figure 10. B.

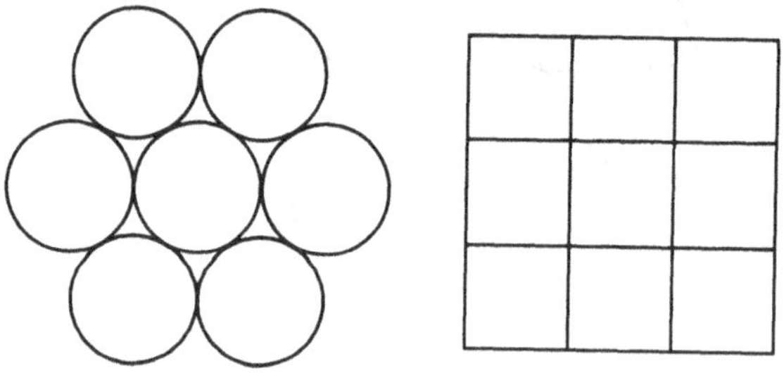

Fig. 10

Grünbaum [149] has proved that $n(A) = 7$ whenever A is not a parallelogram; thus the parallelograms P are characterized by the equality $n(P) = 9$.

The next result provides a good illustration of the close connections among various groups of theorems and methods of proof, since it is of Helly type, but in its simplest form follows immediately from the covering theorem Proposition **34**. Picturesquely stated, it is this: If each two members of a system of congruent disks can be pierced by a needle, then three needles suffice to pierce all the disks of the system. The exact statement is as follows.

44. If a system of congruent disks is such that each two of its members have a common point, then there exist three points such that each disk of the system covers at least one of the three points.

That the "piercing number" $n = 3$ cannot be reduced is shown by the nine disks illustrated in Figure 11.

Fig. 11

According to a conjecture of T. Gallai, there is a theorem of the same sort even when the circles need not be congruent; of course, for this, the piercing number n must be suitably increased. As L. Fejes Tóth reported, ([22], page 97), this conjecture was established by P. Ungár and G. Szekeres for $n = 7$. The sufficiency of $n = 6$ was demonstrated by A. Heppes and of $n = 5$ by L. Szachtó. L. Fejes Tóth conjectured that even $n = 4$ was enough, which was established in an unpublished work of L. Danzer.

We close this section with a series of propositions that treat covering by arbitrary closed sets and that are among the fundamental theorems of point-set topology. A closed set is called *connected* if it cannot be decomposed into two disjoint nonempty closed subsets. Such a closed connected set is also called a *continuum*. Segments, disks, and circles are the simplest examples of continua. A useful lemma on covering follows immediately from these definitions.

45. If a continuum C is covered by finitely many closed sets $A_i (i = 1, \cdots, n)$, but not by any one of them, then there is a point of the continuum that belongs to at least two of the sets A_i.

Essentially deeper is a lemma of E. Sperner [97], which can be formulated as follows for the plane.

46. If an equilateral triangle of side 2 is covered by three closed sets A_i $(i = 1,2,3)$ of diameters less than $\sqrt{3}$, then some point of the triangle belongs to all three of the sets A_i.

This lemma provides the simplest approach to the Lebesgue-Brouwer tiling theorem, which is important in topology as characterizing the number of dimensions, and to which we must refer occasionally in the following sections. The essence of this theorem is contained in the following simple assertion.

47. If an equilateral triangle of side 2 is covered by finitely many closed sets of diameter less than 1, then some point of the triangle belongs to at least three of the sets.

6. Point Set Geometry and Convexity

We shall now consider some set-theoretic contributions to the geometry of ovals.

A set A is said to be *starshaped* with respect to the point P of A, starshaped from P, if the intersection of A with each ray emanating from P is a segment or is the entire ray. Obviously A is starshaped from P if and only if for each point Q of A the entire segment PQ lies in A.

48. The set of all points from which a bounded closed set is starshaped is empty or is an oval.

For the especially simple case of a polygonal domain, we formulate a proposition on starshapedness that is related to Helly's theorem.

49. If a domain is bounded by a simple closed polygon, and for each three sides of the polygon there is an interior point from which these three sides are visible, then there is some point of the domain from which all the sides are visible.

Thus we have a characterization of starshapedness for polygonal domains. (See Figure 12.) For this theorem, discovered by M. Krasnosselsky [65] and sharpened by G. Hajos (see J. Molnar [74]), the following picturesque form is given by I. M. Yaglom—W. G. Boltyanskiĭ [114], problem 20: If for each three paintings in a gallery, one can find a place from which all three can be viewed, then there must be a place in the gallery from which all its pictures are visible.

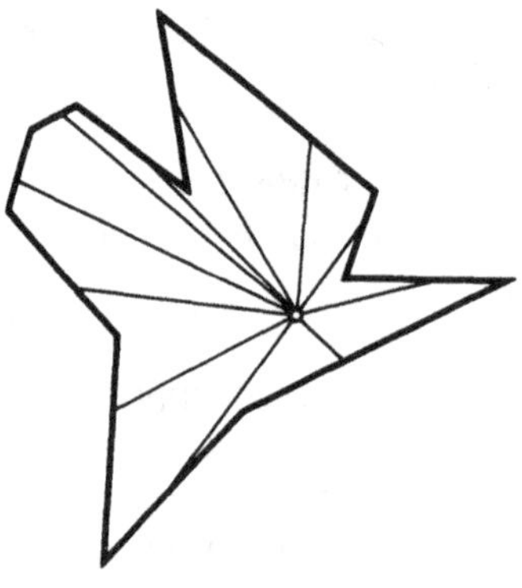

Fig. 12

A set A will be called *locally convex* at one of its points P if there is a neighborhood of P, that is, a disk of positive radius centered at P, such that for each two points Q and Q' of A that belong to this neighborhood, the entire segment QQ' is also contained in A. Under certain restrictions one can derive convexity in the large from convexity in the small.

50. A bounded continuum that is locally convex at each point must be an oval.

By means of theorems of this sort, the usual characterization of convexity can be essentially weakened. Occasionally this is useful, as with the following interesting variant of the notion of convexity, due to F. A. Valentine [105].

51. If a bounded closed set includes with each three of its points at least one of the three segments that they determine, then it can be represented as the union of three or fewer ovals.

7. Realization of Distances

The following propositions treat the distances realized by the pairs of points of a set.

We begin with an estimate for the number of pairs that realize the diameter of a finite set.

52. The greatest distance arising from a set of n points is realized by at most n different pairs of points.

A short proof of this result, to which we have already alluded in connection with Proposition **35**, is given by P. Erdös [18]; see also a problem of H. Hopf–E. Pannwitz [48]. That there are sets of n points in which the greatest distance is realized n times is shown by the set consisting of a point P and $n - 1$ additional points Q_i on a circle with center P and radius 1 such that the greatest distance among the Q_i's is equal to 1. In a problem, P. Erdös [20] has further indicated that in a set of $3n$ points whose diameter is 1, distances of at least $1/\sqrt{2}$ can be realized by at most $3n^2$ pairs of points. This assertion cannot be sharpened through the replacement of $1/\sqrt{2}$ by $1 - \epsilon$ nor by lowering of the bound $3n^2$; this is shown by the set of vertices of n concentric equilateral triangles, similarly situated and with lengths of sides increasing from $1 - \epsilon$ to 1.

Questions as to which distances are realized are answered by theorems of the Steinhaus-Rademacher type, whose hypotheses belong to measure theory. These cannot be considered in the framework of combinatorial geometry. However, the following treatment of the material shows that one can formulate analogous theorems in which the measure-theoretic assumptions are replaced by those of purely set-theoretic character.

We first consider circles and circular disks.

53. If the circle is covered by three closed sets, at least one of the three includes two points at angular distance $a = 2\pi/3$. This assertion is no longer true when $a \neq 2\pi/3$.

54. If the circle is covered by two closed sets, at least one of the two is such that its pairs of points realize all angular distances a of the interval $0 < a \leqq \pi$.

The assumption of closedness of the covering sets is unavoidable. For example, if the circle is decomposed into two disjoint congruent half-closed semicircles, then neither of these includes two points at distance π.

55. If a disk of diameter 1 is covered by two closed sets, at least one of these is such that its pairs of points realize all distances d of the interval $0 < d \leqq 1$.

This implies, in particular, the impossibility of decomposing a disk into two sets, both of smaller diameter than the disk. This is the plane case of the theorem of K. Borsuk [8], according to which a k-dimensional solid sphere cannot be decomposed into k subsets of smaller diameter. To confirm this, one notices that the decomposition into subsets induces a covering by the closures of these subsets, and that the diameter of a set is equal to that of its closure.

Now we turn to similar questions for segments and lines.

56. If a segment of length 1 is covered by two closed sets, at least one of the sets is such that its pairs of points realize all distances of the interval $0 < d \leqq 1/3$.

It can easily be shown that the given interval of distances cannot be enlarged. In fact, suppose $1/3 < c < 1$, and let the interval $0 \leqq x \leqq 1$ be covered by two point sets A and B, where A is described by the inequalities

$$0 \leqq x \leqq \frac{5 - 3c}{12}, \qquad \frac{7 + 3c}{12} \leqq x \leqq 1$$

and B by the inequalities

$$\frac{5 - 3c}{12} \leqq x \leqq \frac{7 + 3c}{12}.$$

No pair of points from A realizes the distance $d' = (7 + 3c)/24$, and none from B realizes the distance $d'' = c$. Since $1/3 < d' < d'' = c$, this shows that neither of the sets yields all distances of the interval $0 < d \leqq c$.

There is an immediate extension of Proposition **56** to the case of the line, which can be regarded here as an infinitely long segment.

57. If the line is covered by two closed sets, at least one of them is such that its pairs of points realize all distances d of the interval $0 < d < \infty$.

Here it is perhaps of interest that a corresponding theorem does not hold for continuous curves that lie in an arbitrarily small neighborhood of a line. Let ϵ be an arbitrary positive number. We consider a parallel strip of breadth 10ϵ and in this, as shown in Figure 13, a periodic polygonal curve with

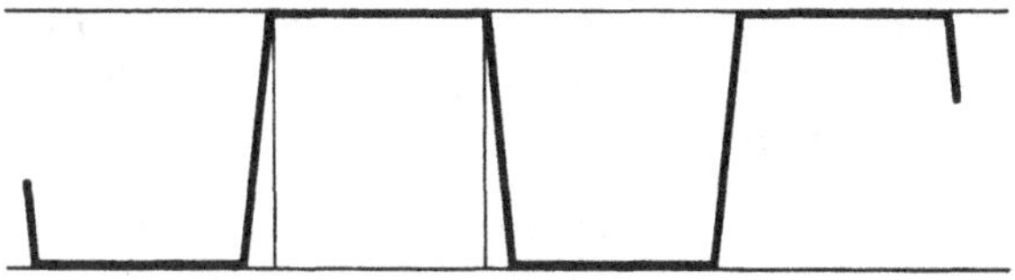

Fig. 13

horizontal segments of length 8ϵ and slanted segments of length $\epsilon\sqrt{101}$. If we cover this continuous curve by the set A of horizontal and the set B of slanted segments, the interval $8\epsilon < a < 10\epsilon$ is lacking in the distance set of A and the interval $\epsilon\sqrt{181} < b < 18\epsilon$ in that of B. Neither of the sets yields all distances.

The segments determined by pairs of points of A are called *chords* of A. Concerning the lengths of chords of plane continua, the following is known.

58. If a continuum A admits a chord S of length $s > 0$, and if $0 < t < s$, then A admits also a chord T of length t. However, this assertion is false if it is required in addition that T should be parallel to S.

The second part of this theorem is verified by the part of the curve $y = \sin x$ for which $0 \leqq x \leqq 2\pi$; while there is exactly one chord S of length 2π, namely on the x axis, there is no chord parallel to S whose length is between π and 2π.

The set of values of t for which there is a chord parallel to S can be much further restricted. We show this by an example due to P. Lévy [72]. Suppose $0 < t < 1$. Let the continuum A be the graph of the function

$$y = f(x) = \sin^2\left(\frac{\pi x}{t}\right) - x\sin^2\left(\frac{\pi}{t}\right),$$

defined in the interval

$$0 \leqq x \leqq 1.$$

Since $f(0) = f(1) = 0$, A admits a chord of length $s = 1$ parallel to the x-axis. On the other hand, $f(x + t) = f(x) - t\sin^2(\pi/t)$, so it follows that $f(x + t) \neq f(x)$ when $t \neq 1/n (n = 1,2,3, \cdots)$. Hence in this case A admits no chord of length t parallel to S. The exceptional values are not peculiar to this example; in fact, the theorem of P. Lévy asserts that each continuum with a chord of length s must admit a parallel chord of length $t = s/n$. The example given shows that there are no other exceptional values. More detailed studies of this and similar questions have been carried through by H. Hopf [46].

Finally, we turn to similar questions that concern coverings of the entire plane.

59. If the plane is covered by three closed sets, at least one of the three is such that its pairs of points realize all distances of the interval $0 < d < \infty$.

If one requires in addition that the covering sets shall be congruent to each other, the proposition can be improved in that the same conclusion holds for a larger number of sets.

60. If a closed set is such that one can cover the plane with five sets congruent to it, then its pairs of points realize all distances d of the interval $0 < d < \infty$.

The question naturally arises whether the number $n = 5$ of the last proposition can be replaced by a larger number. With the following example, we verify that the highest conceivable number is $n = 6$, while $n = 7$ is surely impossible.

In order to obtain a covering of the plane by seven mutually congruent sets, we start from a packing of the plane by regular hexagons. One of these hexagons together with its six neighbors forms a symmetric polygon of 18 sides, which itself can be packed in the plane (see Figure 14). If we group

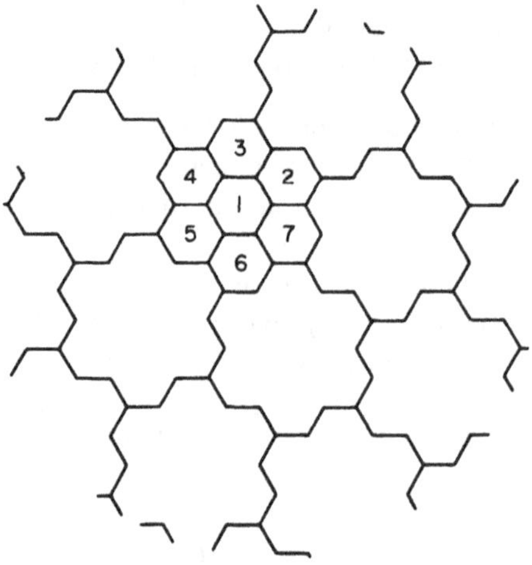

Fig. 14

together in a single set the corresponding hexagons from the various 18-gons, we obtain seven sets that together cover the plane and are translation-equivalent. In the figure, the representative hexagons are marked by the numbers 1 to 7. If the length of a side of the hexagon is $s = 1$, then the pairs of points belonging to one of the seven sets yield all distances d except for those of the interval $2 < d < \sqrt{7}$; thus these sets do not possess the property required in Proposition **60**.

The propositions discussed above are the plane cases of more general theorems that can be stated for the k-dimensional space. See H. Hadwiger [31, 32].

Further investigations of these questions are also presented by A. Heppes [44]. Following a suggestion of P. Erdös, he considers decompositions of the space into disjoint subsets in place of covering by closed sets. For example, one can ask whether a theorem corresponding to Proposition **59** can be formulated for the case in which the plane is decomposed into three disjoint subsets. So far as we know, this is still unsettled.

We end the present group of propositions with a remark on the realization of triples of distances.

61. If the plane is covered by two closed sets, each triangle can be placed so that its vertices are all in the same set.

8. Simple Paradoxes Involving Point Sets

The theory of point sets has produced a series of results that stand in contradiction to our natural feelings and naive geometrical intuition, and which in this sense are often described as paradoxes. However, these are merely unexpected consequences of abstractly precise inferences, revealing in the infiniteness of the continuum a state of affairs that our imagination is no longer able to follow. In most cases rather weak additional assumptions suffice to restrict the domain of possibilities to those with which we are familiar from elementary geometry. There are even very easily verifiable facts of this sort, which can be treated in the realm of an elementary geometry of point sets. The following discussion is concerned with such results.

62. There are bounded sets that are congruent with proper subsets of themselves. For bounded closed sets this possibility does not arise.

The first part of the proposition is verified by means of an example. This will be given here. Let P_0 and Z be two different points, ϕ a rotation about Z through an angle ω that is incommensurable with a full rotation, so that ω/π is irrational. Let A_0 denote the set of all points $P_n = \phi^n(P_0)$ $(n = 0, 1, \cdots)$ that arise from P_0 by n-fold iteration of the rotation ϕ. If we set $A_1 = \phi(A_0)$, then A_1 is congruent with A_0; it is, however, a proper part of A_0, since A_1 does not include the point P_0. Note here that there is no $m > 0$ for which $P_m = P_0$, since otherwise $m\omega$ must be congruent to 0 (mod 2π), which is impossible because of the condition on ω.

This simple example yields a set A_0 including a point P_0 for which $A_0 \simeq A_0 - (P_0)$, where the symbol $\simeq$ indicates congruence and the subtraction denotes deletion of a point from the set. W. Sierpinski [94], page 10, asked whether there is a plane set A such that $A \simeq A - (P)$ for each point $P \in A$. However, this is impossible. In addition, E. G. Straus [101] showed that for any two different points P and Q of a plane set A, the relations $A - (P) \simeq A \simeq A - (Q)$ are impossible.

63. There are unbounded sets that can be decomposed into two disjoint subsets, both congruent with the original set. For bounded sets this possibility does not arise.

We verify the first part of this proposition by an example of S. Mazurkiewicz and W. Sierpinski, see W. Sierpinski [94], page 59: In the complex plane let $\phi(z) = z + 1$ and $\psi(z) = e^i z$, and let C denote the set of all points obtainable from $z = 0$ by means of finite iteration of the transformations given by ϕ and ψ respectively. Let $A = \phi(C)$ and $B = \psi(C)$, so that $C = A \cup B$ and A,B, and C are congruent. Further, A and B have no common point, for otherwise a relation

$$a_0 + a_1 w + \cdots + a_n w^n = 0$$

must subsist, where the $a_\nu(\nu = 0, \cdots, n)$ are integers that are not all zero, and where $w = e^i$. Since e^i is not an algebraic number, this is impossible.

Following J. von Neumann [79], page 85, a point-set A containing more than one point is called *smaller* than the point-set B, written $A < B$, if there is a biunique transformation of A onto B such that for each pair of distinct points $P, Q \in A$ at distance $d[P,Q]$, the condition $d[P,Q] < d[\phi(P),\phi(Q)]$ is satisfied; thus the transformation increases all distances. The following proposition expresses a plausible state of affairs, whose first part is justified by a similarity transformation of the plane onto itself.

64. There are sets A that are smaller than themselves, so that $A < A$. For bounded sets this possibility does not arise.

On the other hand, the following is less plausible.

65. A bounded set A can be smaller than another set B of the same diameter, so that $A < B$ although $D(A) = D(B)$. For bounded closed sets this possibility does not arise.

Again the first part is verified by means of an example. In the complex plane we consider the set A, consisting of the points $z_n = e^{i\pi/3n}(n = 1,2, \cdots)$, and the set B, consisting of $z'_0 = 0$ and $z'_m = e^{i\pi/3m}(m = 1,2, \cdots)$. Obviously $D(A) = D(B) = 1$. The transformation $B = \phi(A)$ is defined by the simple formula $z'_{n-1} = \phi(z_n)$. But then

$$|z_n - z_m| = |z'_n - z'_m| = 2 \sin\left(\frac{n\pi - m\pi}{6mn}\right)$$

for $n > m > 1$ and thus for these indices, $|z_n - z_m| < |z'_{n-1} - z'_{m-1}|$. Further,

$$|z_n - z_1| < |z'_{n-1} - z'_0|.$$

This means that ϕ increases all distances, so that $A < B$. See H. Hadwiger [34].

So far we have always referred to possibilities that one would not have expected in advance; conversely, there are also situations in which a possibility that we should at first expect is, in fact, nonexistent. Here is an example.

66. It is impossible to decompose an oval into two disjoint congruent subsets.

Of course the strength of this assertion lies in the fact that the impossibility is claimed for *all* ovals. Even in special cases the situation had repeatedly drawn interest. Thus B. L. Van der Waerden [108] once posed the problem of proving that it is impossible to decompose a circular disk into two disjoint congruent subsets. There are various simple proofs that are especially suitable for the disk in that they make essential use of the existence of a center of symmetry. However, D. Puppe showed that this fact is not needed, and has very simply proved the impossibility of the required decomposition for ovals whose boundaries contain no segments.

Let us return the center to the field of discussion. On casual consideration, one is inclined to think it impossible to decompose into two disjoint congruent parts a bounded centrally symmetric set that includes its center of symmetry. However, such a decomposition is possible, as is shown by the example of the next paragraph.

In the complex plane, we consider the set A of points

$$z_n = 1 - e^{in} \ (n = 0,1, \cdots)$$

and the set B of points

$$z_{-n} = -1 + e^{in} \ (n = 1,2, \cdots)[i = \sqrt{-1}].$$

A and B are disjoint sets that lie on the circles of radius 1 about the points $+1$ and -1. Both sets are similar to the set A_0 described in connection with Proposition **62**, if the w is set equal to 1. The union $C = A \cup B$ is centrally symmetric and includes the center $z_0 = 0$. The transformation

$$f(z) = -1 + e^i(1 - z)$$

carries the set A onto the set B, since

$$f(z_n) = z_{-(n+1)} \ (n = 0,1, \cdots).$$

But this transformation is a rotation about the point $w = (e^i - 1)/(e^i + 1)$ with the angle of rotation $\theta = 1 + \pi$.

9. Pure Combinatorics. Graphs

We speak of a *partition* or *division into classes* of a set K or a system K if K is the union of disjoint sets K_i, the *classes*. The following group of propositions treats some purely combinatorial questions that arise in connection with partitions of finite or infinite sets and especially also of ordered sets. In combinatorial geometry their role is solely that of useful tools. Although the geometric aspect is in most cases inessential, we shall choose a suitable geometric formulation for the results. Thus our geometric realm will be better accounted for and also a clearer intuitive perception will be attained. In particular, certain combinatorial theorems can profitably be based on graphs (networks). We begin with the following theorem.

67. There exist numbers $N_k(n,p)$, dependent only on the natural numbers k,n,p, for which the following is true: If the set A includes at least $N_k(n,p)$ points, and the p-pointed subsets of A are divided into k classes $K_i(i = 1, \cdots, k)$, then there exist k disjoint, possibly empty, subsets $A_1, \cdots, A_k$ of A having at least n points in all, so that each p-pointed subset of A_i belongs to the class $K_i(i = 1, \cdots, k)$. In particular, for $p = 2$ and $k > 1$ the numbers

$$N_k(n,2) = \frac{2k^{n-1} - k^{n-2} - 1}{k - 1}$$

are of the required sort.

It is a simple matter to extend the result to infinite sets. This gives rise to a theorem named for F. P. Ramsey [89], which we present in a somewhat more special form.

68. If the p-pointed subsets of an infinite set A are divided into k classes, then A contains an infinite set all of whose p-pointed subsets belong to one and the same class.

Additional corollaries are concerned with graphs. A finite *graph* (or network, or 1-complex) is a finite system of points (or *nodes*) in which certain pairs of points are joined by segments (or *edges*). The number of points will here be called the *order* of the graph. Since in our context we treat everything in the plane, we imagine the graph as realized in the plane; any crossings of the segments are to be disregarded. B is called a *subgraph* of the graph A if each node of B is also a node of A and each edge of B is an edge of A. A graph is called *complete* if all its pairs of points are joined.

69. Suppose

$$m \geqq 1,\ k \geqq 2,\ N \geqq \frac{2k^{km-1} - k^{km-2} - 1}{k - 1}.$$

A complete graph of order N whose edges are divided into k classes contains a complete subgraph of order m whose edges belong to one and the same class.

For the combinatorial theorem described here in a graph-theoretic setting, a better bound was given by T. Skolem [95], who showed that

$$\mathcal{N} \geqq \frac{k^{km-k+2} - 1}{k - 1}$$

is sufficient.

For another corollary, we call two graphs *complementary* if they have the same nodes and if a pair of points is joined by an edge in one graph exactly where it is not joined in the other.

70. Suppose $m \geqq 1$ and $\mathcal{N} \geqq 3 \cdot 2^{2m-2} - 1$. Of two complementary graphs of order $\mathcal{N}$, at least one has a complete subgraph of order m.

If $M(m)$ is the smallest natural number such that the above assertion holds for $\mathcal{N} \geqq M(m)$, then

$$M(m) \leqq 3 \cdot 2^{2m-2} - 1.$$

The better bounds

$$2^{m/2} \leqq M(m) \leqq \binom{2m-2}{m-1}$$

are due to P. Erdös—G. Szekeres [21] and P. Erdös [19].

71. Suppose $n \geqq 3$. If the edges of a complete graph of order n are divided into k classes such that any two edges in the same class have a common node, then $k \geqq n - 2$.

This is a simple special case of a theorem of M. Kneser [62]; the bounds given for k cannot be improved, since, as is easily seen by mathematical induction, the edges of a complete graph of order n can be divided in the prescribed manner into $k = n - 2$ classes.

An *even graph* is a graph in which each closed path has an even number of edges, more precisely, one in which a finite sequence of nodes must have an odd number of terms if its successive points are joined by edges of the graph and its last point is the same as its first one. The following theorem gains in generality if we permit two points to be joined by several segments that are individually distinguishable. For pictorial purposes one can replace the segments by arbitrary connecting paths between the nodes.

72. A finite graph is an even graph if and only if its nodes can be divided into two classes such that no two points of the same class are joined by an edge.

If by the *degree* of a node of a graph, one means the number of edges that emanate from this node, then the following proposition can be formulated.

73. The edges of a finite even graph whose nodes all have degree at most k can be divided into k classes such that at no node do edges from the same class meet.

See D. König [63]. Figure 15 shows an even graph in which each point is of degree 4; the numbers 1 to 4 in the figure describe a division of the edges into four classes for which the requirement of the above proposition is satisfied.

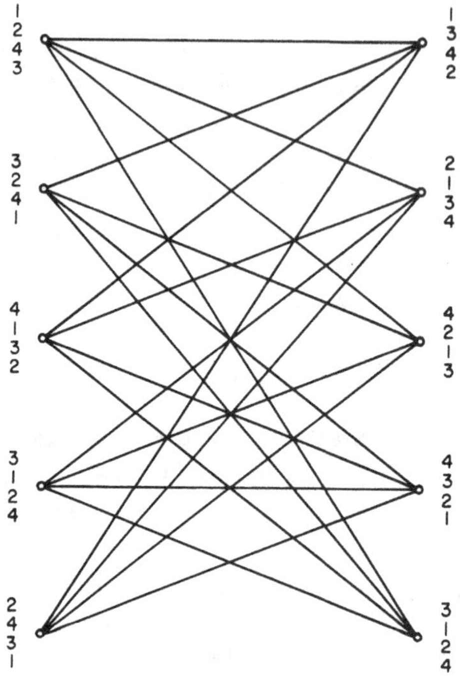

Fig. 15

The following mapping theorem is an important consequence of the theorem on even graphs just discussed.

74. Suppose A and B are finite point sets and k is a natural number $\geqq 2$. If there is a multivalued mapping of A onto B such that each point of A has exactly k image points in B and each point of B has exactly k inverse images in A, then there is a biunique mapping of A onto B for which points are mapped into each other only when they correspond to each other under the multivalued mapping.

More general material on problems of this sort can be found in W. Maak [73]. This well-known combinatorial theorem is easy to remember as the "marriage problem" in the facetious formulation of H. Weyl: In a village, one knows of each boy and girl whether they are friendly or not; what condition is necessary and sufficient for each boy to be able to marry a friendly girl? From the obvious interpretation of the mathematical model of Proposition **74** in relation to the village, it follows that a sufficient but not necessary condition is the existence of a number k such that each boy is acquainted with k girls and each girl with k boys.

The next corollary is of a different type.

75. If a set of $n = pq$ points is divided in two different ways into p classes of q points each, then there are p points such that in both partitions each class includes exactly one of the p points.

According to this assertion, two partitions have a common system of representatives. This was recognized as a purely combinatorial theorem and proved by B. L. van der Waerden [106]. A shorter proof was given by E. Sperner [96].

A theorem on the combinatorics of ordered sets that was especially significant in number theory can be stated in the following geometric terms.

76. Suppose $k \geqq 2$ and $p \geqq 2$. There exist numbers $N(k,p)$ for which the following is true: If the points of a sequence of at least $N(k,p)$ equidistant points on a line are divided into k classes, then there are p equidistant points of the sequence that are all in the same class.

In connection with Baudet's conjecture corresponding to the case $k = 2$, this theorem was first proved in general by B. L. van der Waerden [107]. In arithmetical form, an important corollary asserts that if the natural numbers are divided into k classes and if p is an arbitrary natural number, then at least one of the classes includes p numbers that form an arithmetic progression. It is worthwhile to read the detailed exposition of the discovery of the proof, contained in a series of articles on "Einfall und Überlegung in der Mathematik" ("Inspiration and contemplation in mathematics") by B. L. van der Waerden [109].

Similar theorems can be formulated more generally for lattices. For example, in the unit lattice, the set of all points whose coordinates are integers with respect to an orthogonal coordinate system, the following is known.

77. Suppose $k \geqq 2$. If the points of the unit lattice are divided into k classes and if M is an arbitrary finite set of lattice points, then at least one of the classes contains a subset that is affinely equivalent to M.

This proposition, obtainable from van der Waerden's theorem, is somewhat weaker than a similar theorem of E. Witt [113] that says that at least one class must contain a set that is homothetic to M, that is, similar and similarly situated. Additional material on such questions can be found in A. Khintchine [54] and R. Rado [85].

10. Additional Theorems of Helly Type

In an earlier section we were concerned with some rather simple questions connected with the well-known theorem of Helly. Here in the final section we will discuss in more detail this circle of problems that is characteristic of our subject. We treat several less familiar and more complicated questions of plane combinatorial geometry, those of Helly type predominating.

First we shall see a group of propositions on segments, circular arcs and rectangles.

78. Suppose $p \geqq q \geqq 2$. If a family of at least p segments on a line is such that among each p of the segments, some q segments always have a nonempty intersection, then the segments can be divided into $p - q + 1$ classes so that the segments in each class have a nonempty intersection.

79. Suppose $p \geqq q \geqq 2$. If a family of at least p arcs on a circle is such that among each p of the arcs, some q arcs always have a nonempty intersection, then there are $p - q + 2$ points of the circle such that each of the arcs includes at least one of these points.

80. Suppose $p \geqq q \geqq 2$. There is a smallest natural number $N = N(p,q)$, depending only on p and q, for which the following holds: If a family of at least p mutually parallel rectangles is such that among each p of the rectangles, some q rectangles always have a nonempty intersection, then the rectangles can be divided into N classes so that the rectangles in each class have a nonempty intersection. For this number N the following bounds hold:

$$p - q + 1 \leqq N(p,q) \leqq \frac{(p - q + 1)(p - q + 2)}{2}.$$

The lower bound on N, existence of N at first assumed, is easy to verify. Consider a family consisting of q concentric rectangles and $p - q$ others that have no points in common with each other or with the original q. The hypothesis of the proposition is then satisfied but this family cannot be divided into fewer than $p - q + 1$ classes of the required sort. For the case $p = q = 2$ the bounds yield the result $N(2,2) = 1$, which is merely an alternate statement of the simple Proposition **15**. For $p = 3$, $q = 2$, they show that $2 \leqq N(3,2) \leqq 3$. Actually, $N(3,2) = 3$. This can be seen directly from Figure 16, which shows five rectangles of which each three include a pair with a nonempty intersection.

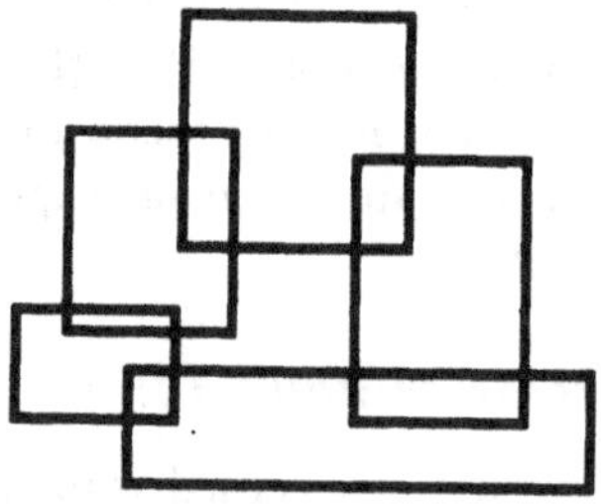

Fig. 16

One verifies similarly that in any case, $N(p,2) \geqq p$ when $p \geqq 2$. If the numbers p and q do not differ too widely, exact values for $N(p,q)$ can be determined.

81. If the supplementary condition $2 \leqq p \leqq q \leqq 2q - 2$ is satisfied, then

$$N(p,q) = p - q + 1.$$

The table below lists the values of $N(p,q)$ in the interval $2 \leqq q \leqq p \leqq 10$, insofar as these are known. In the analogous problem for parallel parallelotopes in the k-dimensional space, the values $\binom{p-q+k}{k}$ constitute upper bounds for $N(p,q)$ if $p \geqq q$; in particular, by Proposition **80** the values $\binom{p-q+2}{2}$ are related in this way to the upper right-hand part of the table.

q \ p	2	3	4	5	6	7	8	9	10
2	1	3							
3		1	2						
4			1	2	3				
5				1	2	3	4		
6					1	2	3	4	5
7						1	2	3	4
8							1	2	3
9								1	2
10									1

Table 1

82. If, in a family of parallel rectangles, there are at most n pairwise disjoint members, then there are $n(n+1)/2$ points such that each rectangle from the family includes at least one of these points.

83. If a finite family of parallel rectangles is such that for each rectangle of the family there are at most m others that are disjoint from it, then there are $m+1$ points such that each rectangle from the family includes at least one of these points.

Now come similar assertions for arbitrary ovals.

84. For certain pairs of natural numbers p and q so that $p \geqq q \geqq 3$, there exist smallest natural numbers $M = M(p,q)$, dependent only on p and q, for which the following is true: If a finite family of ovals has at least p members and is such that among each p members, some q members always have a nonempty intersection, then the ovals can be divided into M classes so that those in the same class have a nonempty intersection. For these numbers M it is true that $p - q + 1 \leqq M(p,q)$.

Comparing the number M defined here with the number N introduced earlier for rectangles, we see that in any case $N(p,q) \leqq M(p,q)$, for each family of rectangles is a family of ovals. Thus lower bounds on M result from the corresponding lower bounds on N. However, while surely the numbers $N(p,q)$ exist $(2 \leqq q \leqq p)$, this is not established in general for $M(p,q)(3 \leqq q \leqq p)$. Thus we lack even an upper bound on the number M. For $p = q = 3$, $M(3,3) = 1$, and this is merely a paraphrase of Helly's theorem stated in Proposition **14**. As in the case of rectangles, the exact values of $M(p,q)$ for arbitrary ovals can be given when p and q are not too different.

85. Under the supplementary condition $3 \leqq q \leqq p \leqq 2q - 3$,

$$M(p,q) = p - q + 1.$$

This is the plane case of the problem treated by H. Hadwiger–H. Debrunner [40]. The table below lists the values of $M(p,q)$ in the interval $3 \leqq q \leqq p \leqq 10$, insofar as they are known. In the cases where the values are unknown, even the existence is uncertain.

q \ p	3	4	5	6	7	8	9	10
3	1							
4		1	2					
5			1	2	3			
6				1	2	3	4	
7					1	2	3	4
8						1	2	3
9							1	2
10								1

Table 2

The simplest case, namely, with the smallest values for p and q, in which the question as to existence and value of the number M is still unsettled is that of $p = 4$ and $q = 3$. From the above results it is apparent only that $M(4,3) \geqq 2$. However, the example of L. Danzer illustrated in Figure 17 shows that $M(4,3) \geqq 3$ if indeed the number exists. Of the six triangles depicted in the figure, each four have a common point. However, the family of triangles cannot be divided into two classes so that those of a class always have nonempty intersection.

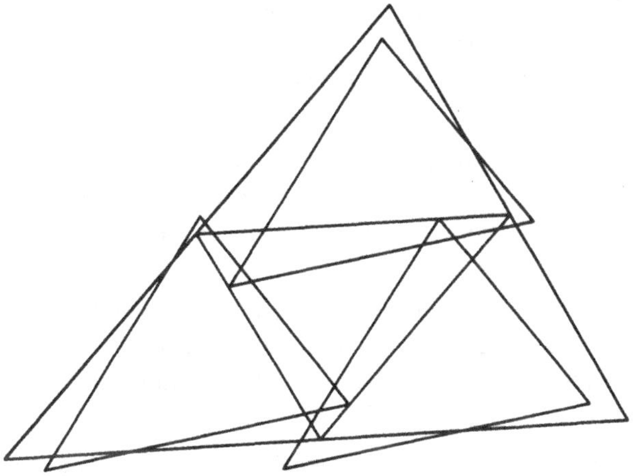

Fig. 17

The fact that the propositions can be sharpened when families of ovals are specialized to families of parallel rectangles is not merely due to technical aspects of the proofs. Just as shown earlier for covering and intersection problems, the rectangles and parallelograms enjoy a special position for Helly type questions also. We shall begin the demonstration with a useful lemma.

86. If a point set A on a circle consists of at least three points, and each three points of A lie in some closed semicircle, then the following alternatives arise: either A is a four-pointed set formed from two antipodal pairs, or A itself lies entirely in a semicircle.

Now the plane case of a theorem of B. Sz.–Nagy [79] can be derived.

87. If, in a family of ovals that are all homothetic to a parallelogram A, each two have a nonempty intersection, then they all have a nonempty intersection. The assertion is no longer true when A is a proper oval that is not a parallelogram.

The assumption that the parallelograms are similar and similarly situated can be weakened, it being necessary only to require that they are all mutually parallel in a sense similar to that employed earlier for rectangles. This is true also of another proposition of the same sort.

88. If a family of ovals all homothetic to a parallelogram A is such that for each line there is a parallel line intersecting all the ovals of the family, then the ovals have a nonempty intersection. The assertion is no longer true when A is a proper oval that is not a parallelogram.

We have already mentioned variants of Helly's theorem in which lines meeting all the ovals of a family, that is, common transversals, are considered in place of points common to all the ovals. Such theorems can be established only if proper assumptions are made concerning the form and distribution of the ovals.

89. For a finite or countably infinite family of disjoint ovals there exists a common transversal if and only if the ovals of the family admit an order relation such that each three ovals of the family are intersected in the given order by a suitably chosen line.

This theorem of H. Hadwiger [39] yields as a special case Proposition **27**, in which a partitioning of the plane by means of strips implies an order relation of the required sort. A theorem of B. Grünbaum [26] can also be derived easily.

90. Suppose that a finite or countably infinite sequence of ovals has the following properties: (a) if $C_n (n = 1,2, \cdots)$ is the convex hull of the first n ovals $A_i (i = 1, \cdots, n)$ and D_n is the convex hull of the remaining ovals A_j $(j = n + 1, n + 2, \cdots)$, then C_n and D_n are disjoint for $n = 1,2, \cdots$; (b) each three ovals of the sequence are intersected by a line. Then there is a line that intersects all the ovals of the sequence.

N. H. Kuiper [66] has established other conditions of this sort; his are somewhat deeper and cannot be formulated in such an elementary fashion.

A special advantage arises when it is assumed in advance that the number of ovals is infinite. The following theorem of H. Hadwiger [37] is known.

91. If each three ovals of an infinite family of pairwise disjoint congruent proper ovals are intersected by some line, then there is a line that intersects all of them.

Here the theorem is no longer true if any one of the four special conditions (proper, congruent, disjoint, infinite) is omitted. (1) The four segments shown in Figure 18 have the property that each three of them can be intersected in inner points by a line; however, no line intersects all four. If one imagines very small rectangles in place of the segments, each containing a countable infinity of disjoint segments of the same length, then again each three of these segments will be intersected by a line, but not all of them.

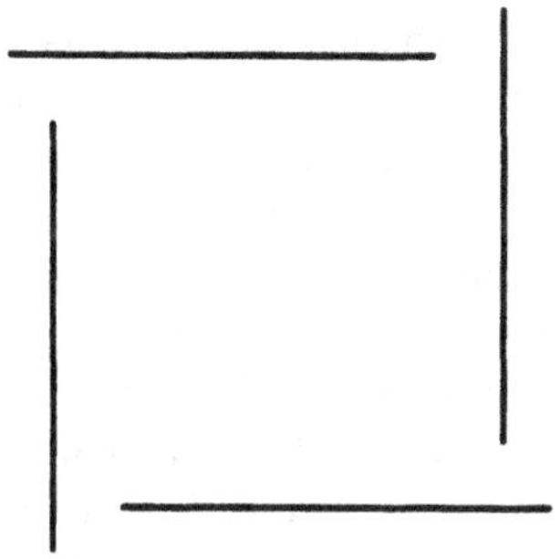

Fig. 18

The ovals are not proper. (2) Of the four circular segments that are bounded by the circumcircle of a square and the sides of the square, we choose a pair of opposite segments. We form the rosette of circular segments that arises when the original pair is rotated through angles of $\pi/4$, $2\pi/4$, $3\pi/4$, and so forth, about the center of the square, and at the same time is stretched out from the center by a factor 2,4,8, and so forth. Each three of the circular segments are then intersected by a line, but not all of them. The ovals are not congruent. (3) If we choose the four vertices of a square of side s and infinitely many points in a sufficiently small neighborhood of one of these vertices as centers of congruent disks whose radii are almost $s/2$, then each three of the disks can be intersected by a line, but not all. The ovals are not disjoint. (4) If, about the five vertices of a regular pentagon, one forms congruent disks whose radii are almost equal to half the length of a side, as shown in Figure 19, then each three, and even each four, of these disks are intersected by some line, but not all five. The number of ovals is not infinite.

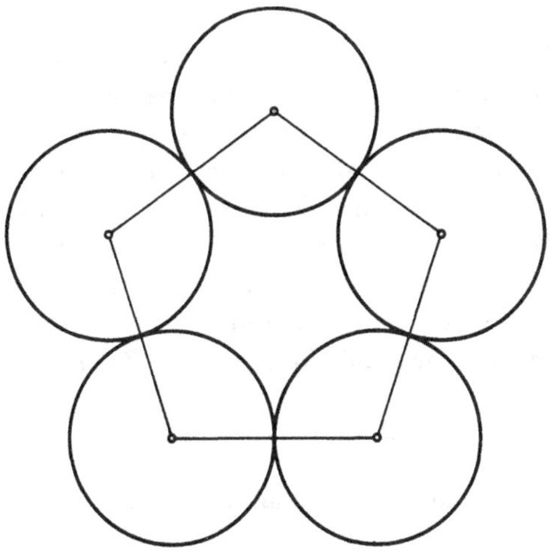

Fig. 19

It is natural to inquire whether more special theorems of the sort discussed here can be formulated for finite families of congruent and disjoint ovals, if supplementary conditions are imposed on the shape of the domains. In particular, one may ask: "Is there for each natural number n a smallest natural number $k = k(n)$ such that n disjoint congruent circular disks always admit a common transversal if each k of them admit a common transversal?" L. Danzer [11] proved that such a "stabbing number" really exists for each n, and that $k(n) \leqq 5$. The family of disks in Figure 19 shows that $k(5) = 5$. Interestingly, the number $n = 5$ is distinguished by being the only one corresponding to Danzer's upper bound $k = 5$, for B. Grünbaum [28] has recently proved that $k(n) = 4$ when $n \geqq 6$. Thus the question posed above is completely answered by the results:

$$k(n) = n \ (1 \leqq n \leqq 5);$$

$$k(n) = 4 \ (5 < n < \infty);$$

$$k(\infty) = 3.$$

The last assertion is a short way of presenting Proposition **91**. Similar results were obtained by B. Grünbaum [28] concerning finite families of disjoint congruent squares or parallelograms. One can omit both the congruence and the limitation on numbers if an assumption is inserted expressing the fact that the disks are sufficiently sparsely distributed.

92. If a family of disks in the plane is so sparsely distributed that even the disks with the same centers but doubled radii are all disjoint, and if each three disks of the family have a common transversal, then there is a transversal common to all.

With the stabbing number four, instead of three, this is the plane case of a problem of H. Hadwiger [38] on k-dimensional spheres. J. Schär noticed that the proposition can also be proved with the stabbing number three.

Conversely, from transversal conditions one can occasionally conclude something about the distribution. For a result of this sort we first formulate a definition. Starting from a countably infinite family of ovals, arbitrarily distributed in the plane, we denote by $N(R)$ the number of ovals in the family that lie entirely inside a disk of radius R about a point of reference Z; then the *density number* of the family for the radius R is given by

$$\Delta(R) = N(R)/\pi R^2.$$

If this density becomes arbitrarily small for large R, that is if

$$\liminf N(R)/R^2 = 0 \text{ for } R \to \infty,$$

then one can say that the family of ovals is more sparsely distributed than under any regular distribution. But if even

$$\liminf N(R)/R = 0 \text{ for } R \to \infty,$$

then we will call the family *extremely sparsely distributed.*

93. For each natural number n the following is true: If an infinite family of proper and mutually congruent ovals is such that no line can intersect more than n ovals of the family, then the family of ovals is extremely sparsely distributed.

The assumption that the ovals should be congruent can also be weakened to the hypothesis that each of the infinitely many ovals contains a disk of fixed radius ρ and can be covered by a disk of fixed radius r. This is true also of the following theorem.

94. If, for an infinite family of proper and mutually congruent ovals, there exists a direction such that each line in this direction meets at least one of the ovals, then there are in the family infinitely many triples of ovals that admit a common transversal.

A most attractive special case is the following situation, illustrated by Figure 20.

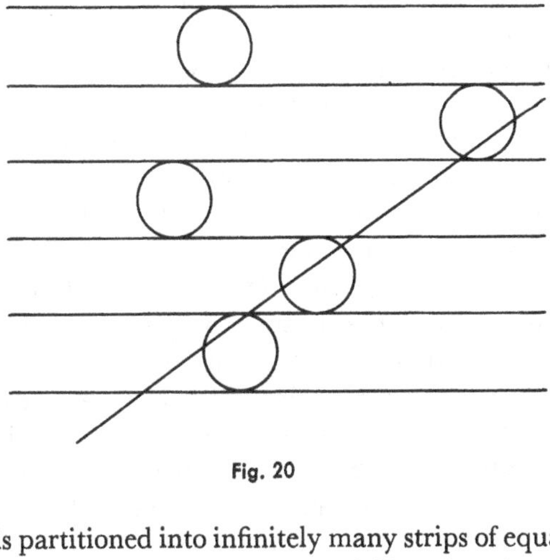

Fig. 20

95. If the plane is partitioned into infinitely many strips of equal breadth by a system of parallel and equidistant lines, and if each strip contains a disk whose diameter is equal to the breadth of the strip, then some three of the disks admit a common transversal.

This last proposition was set as a problem by H. Kneser–W. Süss [61]; E. Hopf gave a solution.

The selection of special subfamilies can also be effected in terms of common points instead of common transversals. Here we have the following propositions, which take us back to the neighborhood of Helly's theorem.

96. If each line meets only finitely many ovals in a given infinite family of ovals, then there is an infinite subfamily consisting of mutually disjoint ovals.

97. If an infinite family of mutually parallel rectangles does not include infinitely many that are pairwise disjoint, then some infinite subfamily has a nonempty intersection.

98. If an infinite family of ovals is such that each of its infinite subfamilies includes three ovals with a nonempty intersection, then some infinite subfamily has a nonempty intersection.

11. Further Development of Combinatorial Geometry*

Being now familiar with several aspects of combinatorial geometry, the reader is ready to consult other books and research papers in the field. The purpose of this chapter is to aid him in that endeavor. Actually, he may choose any of several directions. He may make a more detailed study of combinatorial geometry in the plane. He may consider the 3-dimensional or n-dimensional analogues of some of the material presented earlier, and then proceed to a more extensive study of n-dimensional combinatorial geometry. Finally, he may attack some of the fascinating unsolved problems in which the field abounds, for many of these are easily stated and seem to require a bright idea rather than elaborate mathematical machinery. We attempt here to meet all of these possibilities, as well as to mention literature that is related to material of earlier chapters but was published after the book's German edition had appeared. However, we concentrate on the n-dimensional aspect and on unsolved problems.

Since the present chapter is intended to survey a large volume of material, it includes only a few proofs. After a few words about n-dimensional geometry, we proceed to discuss in turn the various areas that were treated in Chapters 1 to 10. It happens that much of the material of Chapters 3, 4, 5, and 10 has recently been included in its n-dimensional form in a report of L. Danzer — B. Grünbaum — V. Klee [130] (henceforth called DGK), so that material will be treated rather briefly here even though it occupies a large portion of the book. The present chapter is not intended to stand on its own, but rather to serve in conjunction with DGK and other expository material cited below.

*The present chapter reflects the tastes and prejudices of the translator alone. While the authors kindly invited its inclusion, they are not responsible for its contents.

n-Dimensional Geometry

Necessarily, we employ several basic terms from n-dimensional geometry. Most of them are probably familiar to the reader at least in the 3-dimensional case, and he may restrict his attention to that case if he wants to. All of the requisite material can be found in the books of T. Bonnesen — W. Fenchel [121] or H. G. Eggleston [132]. The summary below covers most (but not all) of the notions employed in the subsequent discussion. It will hardly suffice for those readers who have no previous acquaintance with n-dimensional geometry, but for others it may be useful.

We use R^n to denote the n-dimensional space whose points are ordered n-tuples $X = (x_1, \cdots, x_n)$ of real numbers, the *coordinates* of the point. The set of all such points whose coordinates are integers is the *unit lattice* in R^n. *Euclidean* n-space E^n is the space R^n equipped with the *Euclidean distance*

$$d[X,Y] = [\sum_{i=1}^{n} (x_i - y_i)^2]^{1/2}.$$

Topological notions (continuity, closure, interior, boundary) are defined in terms of the distance function, and two subsets of E^n are said to be *congruent* provided there is a biunique distance-preserving correspondence between the points of one and the points of the other. The *diameter* of a subset of E^n is the least upper bound of the distances between its points.

The $(n - 1)$-*sphere* S^{n-1} is the set of all points of E^n that are at unit distance from the origin $(0, \cdots, 0)$, and an *n-ball* of center X and radius r is the set of all points Y of E^n such that $d[X,Y] \leq r$.

For points of R^n, *addition* and *multiplication by scalars* are defined coordinatewise; that is,

$$X + Y = (x_1 + y_1, \cdots, x_n + y_n)$$

and

$$\alpha X = (\alpha x_1, \cdots, \alpha x_n).$$

Also important is the *inner product*

$$X \cdot Y = x_1y_1 + \cdots + x_ny_n.$$

The points X and Y are said to be *orthogonal* provided their inner product is equal to zero.

The *line* determined by two points X and Y of R^n is the set of all points of the form

$$\alpha X + (1 - \alpha)Y \qquad (\alpha \text{ real}),$$

while the *segment* is the set of all points of the form

$$\alpha X + (1 - \alpha)Y \qquad (0 \leq \alpha \leq 1).$$

A subset of R^n is said to be a *flat* provided it contains all of the lines determined by its pairs of points, and to be *convex* provided it contains all of the segments. The *convex hull* of a set A is the intersection of all the convex sets containing it, or equivalently the set of all the *convex combinations*

$$\sum_1^k \alpha_i X_i \qquad \left(\alpha_i \geq 0, \qquad \sum_1^k \alpha_i = 1, \qquad X_i \in A\right).$$

A *linear subspace* of R^n is a flat that includes the origin, or equivalently is a subset of R^n that includes $X + Y$ and αX whenever it includes X and Y and α is real. Every flat is a *translate* of a linear subspace; that is, its points are exactly those obtained by adding a fixed point to the various points of the subspace. The *dimension* of a linear subspace L is the smallest number k such that there exist k points $X_1, \cdots, X_k$ of L whose *linear combinations* $\Sigma_1^k \alpha_i X_i$ (α_i arbitrary) fill the entire set L. The *dimension* of a flat is that of the corresponding linear subspace, and the *dimension* of a convex set is that of the smallest flat containing it.

A *great k-sphere* of S^n is its intersection with a $(k + 1)$-dimensional linear subspace of E^n. Two points X and Y of S^n are *antipodal* provided $X = (-1)Y$. A *k-simplex* in R^n is a k-dimensional set that is the convex hull of $k + 1$ points $X_0, \cdots, X_k$, its *vertices*; the *relative interior* of the simplex is the set of all points of the form $\Sigma_0^k \alpha_i X_i$ with $\Sigma_0^k \alpha_i = 1$ and $\alpha_i > 0$. The k-simplex is *regular* if each two of its vertices determine the same distance.

A *hyperplane* in R^n is a flat of dimension $n - 1$. For each hyperplane H there exists a nonzero point X in R^n and a real number α such that H consists of all points Y for which $X \cdot Y = \alpha$; conversely, every set of this sort is a hyperplane. The hyperplane divides the space into two *open halfspaces*, consisting of the points Y for which $X \cdot Y < \alpha$ and those for which $X \cdot Y > \alpha$. Two sets are said to be *separated* by the hyperplane provided one lies in one halfspace and the other in the other halfspace.

SECTION 1

It seems worthwhile to include an especially short proof of Proposition 1 (Sylvester's theorem), due to L. M. Kelly and reported by H. S. M. Coxeter [10]. Though using more structure than is really needed, it is undoubtedly the proof most readily understood by most readers. Supposing that the points of a finite plane set A are not collinear but nevertheless satisfy the given hypotheses, let the points W, X, and Y of A be such that the distance from W to the line XY is the smallest *positive* number that is realized as the distance from a point of A to a line determined by two points of A. Let Q be the foot on the line XY of the perpendicular from W and let Z be a third point of A that is on the line XY. Then some two of the points X, Y, and Z lie on the same side of Q and a contradiction ensues immediately; for example, if the order is QYZ (where Y may coincide with Q) then the distance from Y to the line WZ is less than the distance from W to the line XY.

Recent papers related to Sylvester's theorem are those of L. M. Kelly—W. O. J. Moser [165], F. Herzog—L. M. Kelly [162], and others cited by these authors. Consider a system consisting of n points in the plane, the t lines determined by these points, and the m lines that contain exactly two of the points. Sylvester's theorem asserts that if $t > 1$ then $m \geq 1$. Kelly and Moser prove that if $t > 1$ then $m \geq 3n/7$, and they show that this result is in a sense the best possible by exhibiting a system in which $n = 7$ and $m = 3$ and another in which $n = 8$ and $m = 4$. However, they conjecture that (with $t > 1$) $m \geq n/2$ when $n \geq 8$ and $m \geq n - 1$ when n is sufficiently large. It was conjectured by P. Erdös [16] and proved by him and others (see [165] for references) that $t > 1$ implies $t \geq n$. Erdös [133] also conjectured that for systems in which at most $n - 2$ points are collinear, $t \geq 2n - 4$ when n is sufficiently large. This is proved by Kelly and Moser for $n \geq 27$ and for $n = 10$; it fails for $7 \leq n \leq 9$ but is conjectured [165] to hold for $10 < n < 27$.

In the theorem of Herzog and Kelly [162], the points in Sylvester's theorem are replaced by sets of points. (For other results in this direction, see papers of B. Grünbaum and M. Edelstein mentioned in [162].) Suppose F is a finite family consisting of at least two bounded closed sets in R^n; suppose the members of F are pairwise disjoint and at least one is infinite. Then some line contains all the members of F or some line intersects exactly two members of F. The theorem is not valid when all the members of F are finite, for the nine points of the Pappus configuration can be divided into three sets such that every line intersecting two of the sets intersects the third one also.

SECTION 2

According to Proposition **5**, the square is the only regular polygon that can be embedded in the unit lattice of E^2. M. S. Klamkin raised the corresponding problem for E^n, and it was solved by H. E. Chrestenson [127] and others. They proved that a regular k-gon can be embedded in the unit lattice of E^n if and only if $n \geq 2$ and $k = 4$ or $n \geq 3$ and $k = 3$ or 6. Thus only the triangle, square and hexagon are embeddable, and they can all be embedded in the unit lattice of E^3. What are the corresponding results for regular polytopes of higher dimension, in particular for the five Platonic solids? (The standard work on regular polytopes is that of H. S. M. Coxeter [128].)

Relative to Proposition **9** and the example following it, we note that for $n \geq 1$ the space E^n contains sets of arbitrarily large finite cardinality that determine exclusively integral distances and do not lie in any hyperplane (F. Steiger [98]). The reasoning of P. Erdös [17] shows that infinite sets of this sort do not exist for $n > 1$, and related results were obtained by R. E. Fullerton [140]. Another reference in E^2 is W. Sierpinski [186].

The following unsolved problem of P. Erdös was reported by H. Hadwiger [154] and S. Ulam [192]: Is there a dense subset of E^2 such that only rational numbers are realized as distances between points of the set? The same question may be asked about E^n with $n \geq 2$, and there appears no obvious relationship between the problem for one value of n and that for another value. As J. R. Isbell has remarked, the possibility suggested by Erdös represents one extreme of what may turn out to be the case. The other extreme possibility, suggested by Isbell, is the existence of an integer $k \geq 5$ such that whenever k points of E^2 determine exclusively rational distances, then some three of the points are collinear or some four are concyclic. This appears to be unsettled even for $k = 5$.

SECTION 3

For a comprehensive report on hull formation, see Section 3 of the report by L. Danzer — B. Grünbaum — V. Klee [130] (DGK). The n-dimensional versions of Propositions **10** and **11** are often known as *Carathéodory's theorem* [126] and *Steinitz's theorem* [99] respectively.*

10_n. If X is a subset of R^n, each point of the convex hull of X is a convex combination of $n + 1$ or fewer points of X.

11_n. If a point is interior to the convex hull of a set X in R^n, it is interior to the convex hull of some set of $2n$ or fewer points of X.

A theorem of W. Bonnice — V. Klee [122] includes both of these results as special cases. For a subset Z of R^n and an integer j between 0 and n, define the *j-interior* of Z, $\text{int}_j Z$, as the set of all points that are relatively interior to some j-dimensional simplex in Z. Then $\text{int}_0 Z = Z$ and $\text{int}_n Z$ is the ordinary interior of Z. They prove that if a point is in the j-interior of the convex hull of a set X in R^n, then it is in the j-interior of the convex hull of some set consisting of max $(2j, n + 1)$ or fewer points of X.

In the n-dimensional version of Kirchberger's theorem [55] (Proposition **12.**), the separating lines are replaced by separating hyperplanes and the four points are replaced by $n + 2$ points. The proof in R^n by means of Helly's theorem (Proposition **14_n**) is very short (H. Rademacher — I. J. Schoenberg [83], also in DGK).

One statement of the n-dimensional version of Proposition **13** is the following, often known as *Radon's theorem* [88].

13_n. Each set of $n + 2$ or more points in R^n can be divided into two disjoint sets whose convex hulls have a common point.

*When k is the number of a theorem stated earlier for the plane, k_n will denote the corresponding n-dimensional result.

Now for each pair of natural numbers n and r, let $f(n,r)$ denote the smallest k such that every set of k or more points in R^n can be divided into r pairwise disjoint sets whose convex hulls have a common point. This function has been studied by R. Rado [86] and B. J. Birch [119] (reported in DGK). They show, in particular, that

$$f(n,r) \geq (n+1)r - n,$$

with equality when $n = 1$, $n = 2$, $r = 1$, or (by Radon's theorem) $r = 2$. Birch conjectures that equality always holds, but even the exact values of $f(3,3)$ and $f(3,4)$ are unknown.

SECTION 4

We now state the n-dimensional form of Proposition **14** (*Helly's theorem*) [42].

14_n. If a family of bounded closed convex sets in R^n is such that each $n+1$ of the sets have a common point, then there is a point common to all sets of the family.

This theorem and its relatives are among the most important in n-dimensional combinatorial geometry. Nevertheless, our discussion here is brief, for we want to avoid unnecessary duplication and they are discussed in detail in DGK.

Most of the intersection theorems for arcs of a circle S^1 have valid analogues in S^n, the unit sphere of E^{n+1}. Four different notions of spherical convexity are employed in various situations, and they are discussed in section 9.1 of DGK. In the least restrictive of these, a subset of S^n is said to be *Horn convex* provided it contains, for each two of its nonantipodal points, at least one of the great circle arcs determined by them. Since the empty set may be regarded as a (-1)-sphere, the following theorem of A. Horn [49] specializes to Proposition **19** when we take $n = 1$ and $k = 2$.

19_n. If a finite family of Horn convex sets in S^n is such that each k of the sets have a common point, where $1 \leq k \leq n+1$, then every great $(n-k)$-sphere in S^n lies in a great $(n-k+1)$-sphere that intersects all the sets of the family.

Propositions **23_n** and **24_n** are quite analogous to this.

With $j = 0$, the common point in the n-dimensional form of Helly's theorem may be regarded as a j-dimensional flat or a j-dimensional convex set that is contained in every set of the family. For $0 \leq j \leq n$, corresponding results have been obtained by R. DeSantis [183] and B. Grünbaum [150], respectively. The common point can also be regarded as a $(j+1)$-pointed set or a j-dimensional flat that intersects every set of the family. Corresponding results for other values of j are mentioned under Section 10 to follow.

SECTION 5

The material of this chapter has extensive ramifications in n-dimensional combinatorial geometry. Most of them are included in sections 6 and 7 of DGK, so here we shall merely describe the higher dimensional results that are most immediately related to those of Section 5.

Proposition **29** extends at once to E^n, with "three" replaced by "$n + 1$." And here are the n-dimensional versions of Jung's theorem [51] and Gale's theorem [24]:

31_n. A set of diameter 1 in E^n can be covered by an n-ball of diameter

$$\sqrt{\frac{2n}{n+1}}.$$

33_n. A set of diameter 1 in E^n can be covered by a regular n-simplex of diameter

$$\sqrt{\frac{n(n+1)}{2}}.$$

For short proofs of sharpened forms of these results, see 2.6 and 6.9 of DGK and other references cited there. For an n-dimensional version of Proposition **34**, see D. Gale [24] and 6.10 of DGK. L. A. Santaló [182] has a spherical analogue of Jung's theorem **31_n**.

A certain duality between **31_n** and **33_n** was noted by D. Gale [24]. The former asserts that if an n-ball is circumscribed about a regular n-simplex of unit diameter, then the ball will cover every set of unit diameter in E^n, while the latter asserts the same covering property for a regular n-simplex that is circumscribed about a ball of unit diameter.

For an extensive survey of matters related to Proposition **35** and especially to the unsolved problem of K. Borsuk [8], see B. Grünbaum [151]. Here we mention only the conjecture of Gale [24] and H. Hadwiger [152] that if d_n is the smallest number α such that an n-ball of unit diameter can be covered by $n + 1$ sets of diameter α, then every set of unit diameter in E^n can be covered by $n + 1$ sets of diameter d_n. This would provide a strong affirmative solution of Borsuk's problem, but it is unsettled even for $n = 3$. (For $n = 2$ it comes from Proposition **35**.) Incidentally, the exact value of d_n seems to be unknown for $n \geq 4$ (Hadwiger [152]). For related details, see Section 5 of Grünbaum [151].

In E^n, a *strip of breadth* 1 is the part of the space between two parallel hyperplanes that are one unit apart. Here is the n-dimensional version of Proposition **38**, established by P. Steinhagen [187] and later simplified by H. Gericke [141]:

38_n. If a convex body in E^n cannot be covered by any strip of breadth 1, it contains an n-ball of diameter

$$\frac{\sqrt{n+2}}{n+1} \quad \text{for even } n$$

$$\frac{1}{\sqrt{n}} \quad \text{for odd } n.$$

For results in E^n that are related to Proposition **42**, see I. C. Gohberg — A. S. Markus [142] and page 137 of DGK. Permitting rotation as well as translation in E^2, D. J. Newman — R. Breusch [176] have proved that a rectangle can be covered by two smaller similar rectangles if and only if it is not a square. For the n-dimensional version of Proposition **43**, see Grünbaum [149], H. Groemer [144], and 7.17 of DGK. For a domain A in E^2, the inequality $n(A) \geq 7$ is established by C. J. A. Halberg, Jr. — E. Levin — E. G. Straus [159] without the assumption of convexity. For results in E^n that are related to Proposition **44** and to the conjecture of T. Gallai, see the discussion under Section 10 to follow.

Proposition **45** is valid in E^n (and more general spaces), and with "finitely" replaced by "countably." In this form, it is due to W. Sierpinski [184]. For expository treatments of Sperner's and other combinatorial lemmas, for their applications in topology, and for additional references see A. W. Tucker [191], H. Hadwiger [156], and H. W. Kuhn [170].

SECTION 6

Propositions **48** and **50** can be extended at once to R^n. The latter is due to H. Tietze and S. Nakajima; it was discussed in a general setting by V. Klee [166], and a still more general result was proved by F. A. Valentine [194]. For additional results and problems related to Proposition **51**, see F. A. Valentine [105, 193].

Here is the n-dimensional version of Krasnosselsky's theorem [65].

49_n. If a bounded closed set in R^n is such that for each $n + 1$ points of the set there is a point from which all $n + 1$ are visible, then the set is starshaped from some point.

SECTION 7

It was conjectured by P. Erdös [18] that if n points in the plane E^2 are the vertices of a convex n-gon, then they determine at least $[n/2]$ different distances, where $[t]$ denotes the greatest integer $\leqq t$. The regular n-gon shows that the number cannot be increased. The conjecture has been proved by

E. Altman [199], who shows also that more than $[n/2]$ distances are determined when n is odd and the n-gon is not regular.

Proposition **52** says that

$$g_2(n) = n,$$

where $g_k(n)$ is the maximum number of times that the greatest distance can occur in a set of n points in E^k. A. Vázsonyi conjectured that

$$g_3(n) = 2n - 2,$$

and this was proved by B. Grünbaum [145], A. Heppes [161], and S. Straszewicz [188]. From the values for $g_2(n)$ and $g_3(n)$ it follows that if $k \leq 3$ and S is a subset of E^k of diameter 1, then some point of S is at unit distance from at most k points of S. This leads to a short inductive proof (Heppes [161]) that every finite subset of E^k is a union of $k + 1$ sets of smaller diameter, a result still unsettled for $k \geq 4$.

H. Lenz noted that

$$g_4(n) \geq \left[\frac{n^2}{4}\right].$$

Indeed, put $s = [n/2]$ and consider the following n points of E^4:

$$(a_i,b_i,0,0), \qquad 1 \leq i \leq s, \qquad (0,0,a_j,b_j), \qquad s + 1 \leq j \leq n,$$

where the numbers a_i,b_i,a_j, and b_j are all strictly between 0 and $1/\sqrt{2}$, and where

$$a_i^2 + b_i^2 = \frac{1}{2} = a_j^2 + b_j^2.$$

These n points form a set of diameter 1, and for each of the $s(n - s) = [n^2/4]$ ways of choosing a pair (i,j), the distance between the points $(a_i,b_i,0,0)$ and $(0,0,a_j,b_j)$ is equal to one. P. Erdös [134] reported Lenz's construction and proved that for every $k \geq 4$,

$$\lim_{n\to\infty} \frac{g_k(n)}{n^2} = \lim_{n\to\infty} \frac{G_k(n)}{n^2} = \frac{1}{2} - \frac{1}{2[k/2]},$$

where $G_k(n)$ is the maximum number of times that the same distance can occur in a set of n points in E^k.

Certain results on covering by closed sets, including some in Section 7, are closely related to theorems on continuous transformations. The connection is expressed in the following observation of K. Borsuk [8].

99. If a subset X of R^m is covered by $n + 1$ closed sets $A_0, \cdots, A_n$, then X can be carried onto a subset Y of R^n by means of a continuous transformation f such that for each point Q of Y, the set $f^{-1}(Q)$ is entirely contained in one of the sets A_i.

Here $f^{-1}(Q)$ is the set of all points P of X for which $f(P) = Q$. Borsuk [8] proved that for every continuous transformation of an n-sphere S^n onto a subset of R^n, some pair of antipodal points in S^n must have the same image in R^n. In conjunction with Proposition **99**, this shows that whenever S^n is covered by $n + 1$ closed sets, at least one of the sets includes two antipodal points.

Various analogues of Borsuk's mapping theorem are related to the following question of B. Knaster [168]:

Suppose $2 \leq k \leq n + 1$, Z is a k-pointed subset of S^n, and f is a continuous transformation of S^n onto a subset of R^{n-k+2}. Must there exist a subset Z' of S^n such that Z' is congruent to Z and all the points of Z' have the same image in R^n?

The answer to Knaster's question is affirmative when $k = 2$ (H. Hopf [163]), when $k = 3$ and $n = 2$ (E. E. Floyd [137]), and in other special circumstances. H. Yamabe—Z. Yujobô [195] settled the case in which $k = n + 1$ and the points of Z are the end points of mutually orthogonal radii; related results for smaller values of k are due to D. G. Bourgin [124]. C. T. Yang [198] established an affirmative answer when $k = 3$ and the points of Z determine an equilateral triangle. The initial stimulus for much of this work was the theorem of S. Kakutani [164], asserting that for every continuous real function on S^2 there exist three mutually orthogonal points at which f has the same value. This implies that every convex body in E^3 admits a circumscribing cube, a result given in E^n by the theorem of Yamabe and Yujobô.

There are many related results that are not covered by Knaster's formulation. For example, Yang [197] shows that whenever f is a continuous transformation of S^{mn} onto a subset of R^m and Z is the set of end points of n given diameters of S^{mn}, then there is a set Z' congruent to Z such that all the points of Z' have the same image in R^m. A theorem of B. A. Rattray [178] implies that if C is a smooth convex body in E^n, centrally symmetric about a point P, then C admits n mutually orthogonal chords through P such that the tangent hyperplanes at their extremities form a rectangular box. For further results and references, see Yang [196], Bourgin [123], M. A. Geraghty [201], H. Debrunner [200], and L. M. Sonneborn [202].

The above mapping theorems have been easy to state, but in many cases their proofs involve a large amount of complicated machinery from both combinatorial and set-theoretic topology. However, in the case of S^2, H. Hadwiger has found elementary proofs of several of the theorems. In particular, his proofs [155, 157] of the following three results will be understandable to most readers of this book.

100. Suppose f is a continuous transformation of S^2 onto a subset of E^2, and U_0 and V_0 are points of S^2. Then some reflection of S^2 on itself carries these points onto points U and V such that

$$f(U) = f(V).$$

101. Suppose f is a continuous real function on S^2, and the points U_0, V_0, and W_0 of S^2 form an equilateral triangle. Then some reflection of S^2 on itself carries these points onto points U, V, and W such that

$$f(U) = f(V) = f(W).$$

102. Suppose f is a continuous real function on S^2, and the points U_0, V_0, W_0, and X_0 are the vertices of a rectangle inscribed in a great circle of S^2. Then some rotation of S^2 carries these points onto points U, V, W, and X such that

$$f(U) = f(V) = f(W) = f(X).$$

These results are related respectively to the more general theorems of H. Hopf [163], C. T. Yang [198], and C. T. Yang [197] stated above, though the first two are sharper in S^2 because of the restriction to reflections. Propositions **101** and **102** were first proved by A. DeMira Fernandes [174] (though not with reflections) and G. R. Livesay [172].

For additional references and for elementary treatments of various mapping and covering theorems related to the above discussion, see H. Hadwiger [155, 156] and A. W. Tucker [191].

SECTION 8

Section 8 was entitled "Simple Paradoxes Involving Point Sets." Now we want to introduce the reader to the most famous of all "paradoxes" in this general area and one of the most striking results in all of mathematics—the so-called *Banach-Tarski paradox*. Its proof would require a degree of sophistication beyond the level of this book, but the reader will surely be able to understand the statement of the result.

Two sets X and Y in E^2 or E^3 are said to be *equivalent by finite decomposition* provided they can be decomposed into the same finite number of sets in such a way that corresponding sets are congruent, or, in other words, provided for some k there exist pairwise disjoint sets $X_1, \cdots, X_k$ and pairwise disjoint sets $Y_1, \cdots, Y_k$ such that

$$X = X_1 \cup X_2 \cup \cdots \cup X_k, \qquad Y = Y_1 \cup Y_2 \cup \cdots \cup Y_k,$$

and

$$X_i \simeq Y_i \qquad (i = 1, \cdots, k).$$

The simplest form of the paradox (S. Banach—A. Tarski [116]) is the following.

103. If S is a closed solid sphere in E^3, or is the surface of such a sphere, and T is the union of two disjoint sets congruent to S, then S and T are equivalent by finite decomposition.

R. M. Robinson [180] obtained such decompositions with $k = 4$ for the surface and $k = 5$ for the solid sphere. In other words, it is possible mathematically to decompose a ball in E^3 into five pieces, then to reassemble three of the pieces to make a congruent copy of the ball and to reassemble the other two pieces to make a second congruent copy of the ball. This seems fantastic, and is "obviously false," as the "reasoning" of the next paragraph shows.

Suppose X, Y, and the X_i's and Y_i's are as described above, and let μ denote the volume function in E^3. Then

$$\mu(X) = \mu(X_1) + \mu(X_2) + \cdots + \mu(X_k)$$
$$= \mu(Y_1) + \mu(Y_2) + \cdots + \mu(Y_k) = \mu(Y),$$

where the first and third equality follow from additivity of the volume μ (for the X_i's are pairwise disjoint, as are the Y_i's) and the second equality comes from the fact that volumes are preserved under congruence (for $X_i \simeq Y_i$). Since the ball S has only half the volume of the set T, the two surely cannot be equivalent by finite decomposition.

The following three assumptions are involved in the above "proof":

(a) the volume μ is defined for all of the sets involved in the discussion;

(b) $\mu(A \cup B) = \mu(A) + \mu(B)$ whenever all three numbers are defined and the sets A and B are disjoint;

(c) μ is invariant under congruence.

These assumptions are valid if attention is restricted to "nice" sets such as geometric figures, but the definition of finite decomposition did not restrict the nature of the sets involved. Thus the situation here, as in Section 8, is that the sets of the paradoxical decomposition are so complicated that our intuition is quite unreliable in dealing with them. The apparent paradox merely demonstrates the impossibility of defining a notion of volume in E^3 that has properties (b) and (c), gives positive volume to some ball, and is defined for all bounded subsets of E^3. Complete divorcement of the decomposition from any geometric properties is indicated by the following stronger form, also from [116].

104. If X and Y are any two subsets of E^3 that are both bounded and both have interior points, then X and Y are equivalent by finite decomposition.

The above decomposition theorems are far removed from physical reality, not only because of the complex nature of the sets involved but also because the congruence involved amounts to instantaneous transfer of the set from one part of the space to another. Thus the following unsolved problem of J. deGroot is of interest (reported by T. J. Dekker [131]):

Is it possible to decompose a solid sphere S into a finite number of pieces that may be physically moved *without penetrating each other* so that two copies of S are formed? Here a physical motion is a rigid motion continuously dependent on a time-parameter.

As was observed by S. Banach [115], A. P. Morse [175], and others, the Euclidean plane E^2 does not support an analogue of the Banach-Tarski paradox. This follows from the fact that the ordinary notion of area for geometric figures in the plane can be extended to apply to all bounded sets and to satisfy conditions (b) and (c) above. Thus two geometric figures in E^2 that are equivalent by finite decomposition must have the same area. It is unknown, however, whether equality of areas is sufficient in general, though it is sufficient for polygons (A. Tarski [189]). Even the following modern form of the problem of "squaring the circle," proposed by A. Tarski [190] in 1925 is still unsolved:

Are a circular disk and a square of equal area equivalent by finite decomposition?

For further discussion of paradoxical decompositions and for additional references, see W. Sierpinski [94, 185], T. J. Dekker [131], and H. Meschkowski [173].

There is also an interesting (nonparadoxical) theory concerning the decomposition of polygons and polytopes into polygons and polytopes, and treating the notion of equivalence with respect to such decompositions. For expositions of this theory and for additional references, see H. Hadwiger [153], V. G. Boltyanskiĭ [120], and H. Meschkowski [173].

SECTION 9

The theory of graphs and the study of combinatorics have progressed greatly in recent years, partly under the impetus of industrial applications. Among books of probable interest to the reader are those by D. König [169], C. Berge [117], O. Ore [177], L. R. Ford, Jr. — D. R. Fulkerson [138], and C. Berge — A. Ghouila-Houri [118] on graphs, J. Riordan [179] and H. J. Ryser [181] on combinatorics. The discussion below is confined to some material directly related to combinatorial geometry, beginning with a geometrical application in R^n of the finite form of Ramsey's theorem (Proposition **67**). For $n = 2$ this is due to P. Erdös — G. Szekeres [21]; it is discussed also by Ryser [181].

A subset X of R^n is said to be *in general position* provided for $2 \leq j \leq n$, no $(j - 1)$-dimensional flat includes more than j points of X. When $n = 3$, this requires that X should consist of one or two points, of three points not on a line, or of at least four points with no four coplanar. A set Y in R^n is said to be *convexly independent* provided no point of Y is a convex combination of other points of Y. When $n = 3$, this requires that Y should consist of onx or two points or is the set of all vertices of a convex polygon or a convee polytope.

Let $[t]$ denote the greatest integer $\leq t$.

105. If $n + \left[\frac{n+5}{2}\right]$ points of R^n are in general position, then some $n+2$ of them are convexly independent.

106. If a set in R^n is such that each $n+2$ of its points are convexly independent, then the entire set is convexly independent.

Proposition **105** is easily proved in the plane (Erdös — Szekeres [21], Ryser [181]); the n-dimensional proof is similar, though a bit more complicated. Proposition **106** comes at once from Carathéodory's theorem **10_n**.

107. For $n \geq 2$ and $m \geq n+1$, there is a unique smallest number $C(m,n)$ that has the following property: whenever a subset of R^n is in general position and includes at least $C(m,n)$ points, then some m of its points are convexly independent.

We show that, in fact,

$$C(m,n) \leq \mathcal{N}_2\left(m + n + \left[\frac{n+5}{2}\right] - 1, n+2\right),$$

where the function $\mathcal{N}_2$ is as in Proposition **67**. Consider any set in R^n having at least $\mathcal{N}_2(\cdots)$ points, and divide its $(n+2)$-pointed subsets into two classes according as they are convexly independent or not. By the definition of $\mathcal{N}_2$ there exist two subsets A and B of the original set, having together at least $m + n + \left[\frac{n+5}{2}\right] - 1$ points, such that all $(n+2)$-pointed sets from A are convexly dependent and all from B are convexly independent. By proposition **105**, at most $n + \left[\frac{n+5}{2}\right] - 1$ of these points are in A, whence at least m of them are in B. But B is convexly dependent by Proposition **106**, so the proof is complete.

P. Erdös — G. Szekeres [21] proved that $C(5,2) = 9$, and later [136] that $C(r,2) \geq 2^{r-2}$. They conjectured that $C(r,2) = 2^{r-2} + 1$, but this is unsettled. Almost nothing is known about the numbers $C(r,n)$ for $r - 2 > n > 2$.

See H. J. Ryser [181] for additional references on Ramsey's theorem. Some progress in determining the best values for the numbers $\mathcal{N}_k(n,p)$ has been made by R. E. Greenwood — A. M. Gleason [143] and C. Frasnay [139].

The following was established by R. Rado [87], using the infinite form of Ramsey's theorem (Proposition **68**) and the n-dimensional form of Helly's theorem.

108. Each infinite family of convex sets in R^n contains an infinite subfamily F such that

either (i) for some k with $1 \leq k \leq n$ it is true that each k members of F have a nonempty intersection but each $k+1$ members of F have an empty intersection.

or (ii) every finite subfamily of F has a nonempty intersection.

For other intersection properties of infinite families of convex sets, see the report of V. Klee [167].

Now we will discuss two problems that illustrate two of the many contacts between graph theory and combinatorial geometry. The first one asks: What is the smallest number k_n of sets into which the space E^n can be decomposed so that none of the sets includes two points at unit distance? This problem has been considered independently by P. Erdös — H. Hadwiger and by E. Nelson — J. R. Isbell. Clearly $2 = k_1 < k_2 \leq k_3 \leq \cdots$, and it is known also that $4 \leq k_2 \leq 7$ (Hadwiger [158]). The problem may be regarded as one on the coloring of a graph, where the nodes of the graph are just the points of E^n and two nodes P and Q are joined by an edge if and only if $d[P,Q] = 1$. Then the problem asks for the smallest number of colors with which the nodes can be colored so that any two nodes that are joined by an edge have different colors. By a theorem of N. G. deBruijn — P. Erdös [125], a graph can be colored with k colors if the same is true of all of its finite subgraphs. Thus in the problem of determining k_n, it suffices to consider all *finite* subsets of E^n.

Note that however E^n is decomposed into $k_n - 1$ sets, *every* distance is realized in one of the sets, though different distances may require different sets; $k_n - 1$ is the largest integer that has this property. Compare this with some of the problems of Section 7.

Graphs are also useful in describing certain intersection properties of a family F of sets, where we associate a node with every set in the family and join two nodes by an edge if and only if the corresponding sets intersect. The graph G does not describe all of the intersection properties of F, but only those associated with pairwise intersections. However, the one-dimensional form of Helly's theorem (Proposition **16**) shows that G does describe all of F's intersection properties when F is a finite family of segments in the line R^1. The graphs that are obtainable in this way were characterized by C. G. Lekkerkerker — J. C. Boland [171], but the corresponding problem is unsolved for other classes of families of sets — arcs in a circle, circular disks in E^2, convex sets in R^n, and so forth.

SECTION 10

In the terminology of H. Hadwiger — H. Debrunner [40], a *k-partition* of a family F of sets is a way of splitting F into k or fewer subfamilies, each having a nonempty intersection. In Propositions **44** and **78** to **85** we are seeking conditions on a family of convex sets or arcs that will assure its k-partitionability. Though Helly's theorem serves nicely when $k = 1$, various examples show that the situation is much more complex for larger values of k and that sharp results can be expected only under very special circumstances.

For a family F of sets and for natural numbers p and q, let $\mathcal{J}_F(p,q)$ denote the smallest number k that has the following property:

If G is any finite subfamily of F such that for every way of choosing p members from G, some q of those members have a common point, then G is k-partitionable.

When no such k exists, let $\mathcal{J}_F(p,q) = \infty$.

Now suppose F is a family of convex sets in R^n. Then $\mathcal{J}_F(n+1, n+1) = 1$ by Helly's theorem, but F must be severely restricted for $\mathcal{J}_F(k,k)$ to be finite when $k < n + 1$. Results in this direction are Proposition **44** and the conjecture of T. Gallai discussed in Section 5, and the more general theorems of Gallai type that are discussed in Section 7 of DGK. When F is a family of n-balls in E^n, estimates of $\mathcal{J}_F(k,k)$ are given by L. Danzer [129]; the upper bounds on $\mathcal{J}_F(k,k)$ are stronger when the balls in F are all of the same size and thus F is a family of translates of a single ball. Sharper results may follow from a method of P. Erdös — C. A. Rogers [135]. For more general families of translates, the main results are those of B. Grünbaum [146, 147]. It is apparently unknown whether there exists a convex body C in E^n such that $\mathcal{J}_F(2,2) < n + 1$ for some family F of translates of C,

Propositions **78** to **81**, **84** and **85** determine the numbers $\mathcal{J}_F(p,q)$ in various other situations. In the n-dimensional version of Proposition **85** (Hadwiger — Debrunner [40]), the requirement that $3 \leq q \leq p \leq 2q - 3$ is replaced by the condition that

$$n + 1 \leq q \leq p \quad \text{and} \quad (n-1)p + n + 1 \leq nq.$$

Proposition **87** suggests the following definition. For a family F of sets and for a natural number j, let $\beta_j(F)$ denote the largest number l that has the following property:

Whenever a finite subfamily of F is such that each j of its members have a common point, then in fact each l of its members have a common point.

If every l has the property, let $\beta_j(F) = \infty$.

Theorems of B. Sz-Nagy [79] and O. Hanner [160] can be combined as the following generalization of Proposition **87**.

87_n. Suppose C is a convex body in R^n, TC is the family of all translates of C, and HC is the family of all homothetic images of C. Then

$$\beta_2(TC) = \beta_2(HC)\begin{cases} = \infty \text{ if } C \text{ is a parallelotope} \\ \leq 3 \text{ if } C \text{ is not a parallelotope.}\end{cases}$$

Hanner characterizes the special bodies C for which $\beta_2(TC) = 3$. There are none in R^2 but the regular octahedron is one in E^3. See DGK for further details.

Propositions **89** to **96** are concerned with common transversals, a topic discussed also in Section 5 of DGK. Most of the results in this area are restricted to transversal lines (rather then flats of higher dimension) and many of them are known only in the plane. The following statement may be compared with Propositions **89**, **91**, and **92**.

109. Suppose F is a family of bounded closed convex sets in R^n or E^n, and let T_j denote the condition that each j members of F have a common transversal. Then each of the following four conditions implies the existence of a line that intersects all members of F:

(a) $T_{(2n-1)2^n}$ holds and the sets are parallelotopes with axes parallel to the coordinate axes;

(b) T_{2n-1} holds and the sets are situated in distinct parallel hyperplanes;

(c) T_{2n-1} holds and the sets are Euclidean n-balls such that the distance between any two centers is greater than the sum of the corresponding diameters;

(d) T_{n+1} holds, the diameters of the sets have finite upper bound but the union of the sets is unbounded.

The first of these results is due to L. A. Santaló [91]. In the stated forms, **109** (b) and (c) are due to B. Grünbaum [148] and **109** (d) to DGK, but they were first established by other authors in more special form, as is clear from the discussion in Section 10. See DGK for additional results and references.

Part II

Proofs

With the aid of references cited above, the Propositions **1** through **98** will now be justified by short proofs. Often the course of the reasoning will only be outlined due to limitations of space. The arguments are based predominantly on elementary facts, supplemented here and there by simple considerations of point-set geometry.

1. Supposing that the points are not on a line but nevertheless satisfy the hypothesis of the theorem, we reach a contradiction as follows. By a suitable projective transformation, exactly one of the points is carried onto a point P_0 at infinity, one of the "ideal points" added to the affine plane in constructing the projective plane, while the other points are carried into points $P_1, \cdots, P_n$ of the original affine plane. The system of lines determined by the original pairs of points is carried into a system of parallel lines through P_0, each including at least two of the points $P_i (i \geqq 1)$, and a finite system of transversals each including at least three of the points $P_i (i \geqq 1)$. Let G be one of the transversals that forms the smallest angle with the parallels through P_0, and let P_i, P_j, P_k be three points among $P_1, \cdots P_n$ that lie on G in the order indicated. The line through P_0 and P_j includes a third point P_m of the set. But then either the line through P_i and P_m or the one through P_k and P_m forms a smaller angle with the parallels than does G, which contradicts the construction.

2. Proposition **2** is dual to Proposition **1**.

3. Proposition **3** follows as a corollary from Proposition **1**, if by means of inversion in a circle centered at one of the points of the set, the circles through this point are carried into lines that satisfy the hypothesis of Proposition **1**.

4. Let D be the smallest circular disk that covers the set. In the circumference of D, no semicircular arc is free of points of the set; in particular, the circumference includes a point P of the set. But then no point Q of the set can be interior to D, for otherwise reflection of the set in the axis of symmetry of P and Q would produce points of the set that are not in D. Now suppose, in addition to the hypotheses of the theorem, that the set in question is finite and includes more than two points. Let ϕ be the smallest angle between any two different axes of symmetry for the set. Reflection in these two axes amounts to a rotation through an angle of 2ϕ so that the set has corresponding rotational symmetry. Since the regular n-gons with central angle $\phi = 2\pi/n$ are the only sets with these properties of rotational and reflectional symmetry, we see that each finite set with the property described in Proposition **4** is the set of vertices of a regular polygon.

5. If there is a regular n-gon embedded in the lattice (n fixed), then there is one of smallest side, since here the only relevant values are $\sqrt{p^2 + q^2}$ (p,q integers). Assuming the existence, let $P_1, P_2, \cdots, P_n$ be the vertices, in natural sequence, of a smallest regular lattice n-gon. If one translates these lattice points in the given order by means of the lattice vectors $P_2P_3, P_3P_4, \cdots, P_1P_2$, respectively, new lattice points are obtained. For $n = 5$ and $n \geqq 7$ these form a smaller regular lattice n-gon, contradicting the condition of minimality. For $n = 3$, one sees in the following way the impossibility of a regular n-gon embedded in the lattice. The area $s^2\sqrt{3}/4$ would be an irrational number since s^2 is an integer, but on the other hand, by the determinant formula, it would be a rational number. The same reasoning applies to a regular hexagon of area $3s^2\sqrt{3}/2$.

6. The area $s^2 \sin \alpha$ of a lattice rhombus is an integer by the determinant formula. Thus, by Proposition **8**, $\alpha = \pi/6$ or $\alpha = \pi/2$. The first possibility is ruled out since the rhombus is carried into another lattice rhombus by rotation through the angle $\pi/2$ about a vertex, which carries each lattice point into a lattice point. But then a regular lattice triangle is obtained, contradicting Proposition **5**.

7. Proposition **7** is a simple consequence of Proposition **8**.

8. Note that for $n = 5$ and $n \geqq 7$, the reasoning of the proof of Proposition **5** applies more generally in every rectangular lattice. Proposition **8** can be derived from this sharper result that triangles, squares, and hexagons are the only regular polygons that can be embedded in a rectangular lattice. Indeed, suppose $\alpha = (m/n)2\pi$ and the fraction m/n is irreducible. If $\cos \alpha$ is rational, then by trigonometric formulae $\cos \nu\alpha = a_\nu$ and $\sin \nu\alpha = b_\nu \sin \alpha$ with rational $a_\nu, b_\nu (\nu = 1,2, \cdots, n)$. Let N be the least common denominator of the $2n$ fractions a_ν, b_ν. Let a rectangle of length $1/N$ and width $(\sin \alpha)/N$ be placed in the Cartesian plane with one vertex at the origin and long side along the x axis, and consider the rectangular lattice generated by this rectangle. The points at unit distance from the origin $(0,0)$ and having angle of inclination $\nu\alpha (\nu = 1, \cdots, n)$ must all be lattice points, as follows from the choice of N. Since $\alpha = (m/n)2\pi$, these points form a regular n-gon. As was mentioned at the outset, this implies that n has one of the values 1,2,3,4,6, and with the supplementary condition $0 < \alpha < \pi/2$ it follows that $\alpha = \pi/3$.

9. If a point set is given with all its points at integral distances from each other and including three points A,B, and C not on a line, and if k is the larger of the distances $d[A,B],d[B,C]$, then there are at most $4(k+1)^2$, hence only finitely many, points P such that the numbers $d[P,A] - d[P,B]$ and $d[P,B] - d[P,C]$ are both integers. In fact $|\, d[P,A] - d[P,B] \,| < d[A,B]$ and thus can assume only the values $0,1, \cdots ,k$, whence P lies on one of $k+1$ hyperbolas having A and B as foci. In the same way P lies on one of $k+1$ hyperbolas having B and C as foci. There are at most $4(k+1)^2$ points that lie on two of these $2k+2$ hyperbolas.

10. The "if" part is trivial. The "only if" part is clear for a finite set, since its convex hull is a convex polygon whose vertices belong to the set; if this polygon is triangulated by means of segments joining a particular vertex to all the others, each of its points lies in one of the resulting triangles and hence in the convex hull of three or fewer points of the set. It remains to show that for an infinite set M, the convex hull $\bar{M}$ of M is contained in the set N of all points that lie in the convex hulls of finite subsets of M. But it is easily verified that N contains, with each two of its points, the entire line segment joining them; further, N includes each point of M. Since $\bar{M}$ was defined as the smallest set with these two properties, the proof is complete.

11. The "if" part is trivial and we consider the "only if" part. An interior point P of the convex hull $\bar{M}$ of M is interior to a triangle with vertices in $\bar{M}$. By Proposition **10**, each of these vertices is in the convex hull of three or fewer points of M, and hence the entire triangle lies in the convex hull of finitely many points of M. If the convex polygon formed by these points is suitably triangulated, then P is interior to one of the triangles or to the union of two adjacent triangles, hence interior to the convex hull of three or four points of M.

12. The "only if" part is trivial. It remains to show that if two sets M and N are nonseparable, then they contain nonseparable subsets M' and N' that together include at most four points. Now M and N are nonseparable exactly when their convex hulls $\bar{M}$ and $\bar{N}$ have a common point P. For such a P there are (by Proposition **10**) two sets M'' and N'', each including at most three points, whose convex hulls $\bar{M}''$ and $\bar{N}''$ have P in common. Either one of these convex hulls is contained in the other, say $\bar{M}''$ in $\bar{N}''$, or the triangles $\bar{M}''$ and $\bar{N}''$ have crossing edges. In the first case let M' consist of a point of M'',$N' = N''$; in the second case let M' and N' each consist of the two end points of one edge from the crossing pair. In both cases M' and N' are nonseparable, for $\bar{M}'$ and $\bar{N}'$ have common points.

13. Choose four points from the given set M. If their convex hull is a triangle, then one point N is in the convex hull of the other three points, hence surely in the convex hull of $M - (N)$, and the two disjoint sets (N) and $M - (N)$

are not separable. On the other hand, if the convex hull is a nondegenerate quadrilateral, let N consist of the end points of a diagonal. Again N and $M - N$ are disjoint nonseparable subsets of M.

14. For finitely many ovals, Helly's theorem follows through mathematical induction from the following lemma:

Suppose $k \geqq 4$. If each $k - 1$ of k ovals have common points, then all k ovals have a common point.

Proof. Let $C_1, \cdots, C_k$ be the k ovals and let P_i denote a point that is in all except perhaps C_i. By Proposition **13** the points $P_i(i = 1, \cdots, k)$ can be divided into two disjoint classes $M' = \{P_{i1} \cdots, P_{i_m}\}$ and $M'' = \{P_{j_1}, \cdots, P_{j_n}\}$ whose convex hulls $\bar{M}'$ and $\bar{M}''$ have a common point P. But each point of M' belongs to all the ovals except perhaps $C_{i1}, \cdots, C_{i_m}$, and by convexity of the C_i's the same is true of M'; similarly, each point of $\bar{M}''$ belongs to all the ovals except perhaps $C_{j1}, \cdots, C_{j_n}$. The point P belongs to both $\bar{M}'$ and $\bar{M}''$, hence to all the ovals without exception.

If there is no point common to all the ovals of an infinite system, then for each point of an oval C_1 of the system there is another oval C_i of the system that misses this point and hence misses also an entire disk centered at the point. Let us associate C_i with this disk. By the Heine-Borel theorem, C_1 is covered by finitely many of these disks. By the construction, the associated ovals C_i form together with C_1 a finite system of ovals having no common point, contradicting the above result that finitely many ovals of the system do have a common point when the hypotheses of Proposition **14** are satisfied.

15. Proposition **15** follows from Proposition **14** if one knows that three (parallel) rectangles R_1, R_2, R_3 always have a common point when this is so for each two of them. In a Cartesian coordinate system whose axes are parallel to the sides of the rectangles, let $P_i(x_i, y_i)$ be a point that is in all three of the rectangles except perhaps R_i. Note that not only are P_i and P_j points of R_k, but in fact R_k contains the entire rectangle whose sides are parallel to the axes and whose diagonal is the segment P_iP_j; that is, R_k includes all points $P(x,y)$ for which x lies in the interval (x_i, x_j) and y in the interval (y_i, y_j). If the indices are chosen so that $x_1 \leqq x_2 \leqq x_3$ and $y_i \leqq y_j \leqq y_k$, then the point $P(x_2, y_j)$ satisfies these conditions for all three of the rectangles and hence belongs to all of them.

16. Proposition **16** is a corollary of Proposition **15**, since the rectangles can degenerate to segments.

17. Proposition **17** can be reduced to Proposition **14**. A family of circular arcs, each smaller than a semicircle, has a common point if and only if this is true of the corresponding segments of the disk. For this it suffices according to Proposition **14** that each three should have a common point.

18. Proposition **18** follows from Proposition **16**. If each of the circular arcs is smaller than one-third of a circle while each two of them have a common point, then they leave some point of the circle uncovered, for example, the point antipodal to the mid-point of one of the arcs. The circle can be cut at this point and unrolled on a line so that each of the arcs turns into a segment of the line.

19. Let $L(\alpha)$ be the *directed* line through the center Z of the circle, making an angle α with a fixed direction. If the given pairwise intersecting arcs are projected orthogonally onto $L(\alpha)$, the resulting segments have the same property. Thus the intersection $D(\alpha)$ of all of these segments is a point or a segment, in any case not empty (Proposition **16**). For at least one angle α_0 the set $D(\alpha)$ includes the center Z. To see this, note that the position of $D(\alpha)$ relative to Z in $L(\alpha)$ is exactly antipodal to the position of $D(\alpha + \pi)$ relative to Z in $L(\alpha + \pi)$. (Remember that these are directed lines!) Since the orthogonal projection of each arc on $L(\alpha)$ varies continuously with α, so does the set $D(\alpha)$, and thus rotation through an angle of π must yield at least one α_0 for which $Z \epsilon D(\alpha_0)$. The line $L(\alpha_0 + \pi/2)$ is then a diametral line that intersects all the arcs.

The variant Propositions **20** through **28** are derived from the basic results of Propositions **14**, **16**, **17** and **19** by the use of suitable transformations.

20–22. Let us choose a point P of the oval A; the position of a translate of A is completely determined by the corresponding position of P. It is easily verified that if A takes on all positions in which it is contained in an oval B, then the corresponding positions of P form an oval B^*. The same is true of all positions in which A intersects, or contains, an oval B. In this way, each oval B of the family in question is associated with another oval B^*, and the Propositions **20** through **22** are reduced to Proposition **14**.

23. If the pairwise intersecting ovals are mapped into a circle by central projection, they give rise to arcs that satisfy Proposition **19**. Then every oval in the family is intersected by the line determined by the two antipodal points specified in Proposition **19**.

24. By orthogonal projection of the ovals into a line, a family of segments satisfying Proposition **16** is generated. All the ovals of the family are intersected by the perpendicular line that passes through a point common to all these segments.

25. The conclusion is evident if among the parallel rectangles of the family there are two that have a unique ascending transversal in common, for then this line must intersect every other rectangle of the family. Thus we may then assume that each three rectangles of the family admit a common ascending transversal that is not parallel to the x axis. But then the same is true for any finite number of rectangles in the family. To see this, we lay out two lines parallel to the x axis and associate with each transversal a point of an auxiliary plane, the coordinates of this point being the x coordinates of the intersections of the transversal with the two parallel lines. The set of all ascending transversals of a rectangle of the family is thus associated with a convex, closed, but unbounded point set in the auxiliary plane. By hypothesis, each three of these sets have a common point. For any finite number of these convex sets, the intersections with a sufficiently large disk are ovals that according to Proposition **14** have a common point. The line associated with this point intersects the corresponding finite number of rectangles. In order to carry out the proof for infinite sets of rectangles also, without using a stronger variant of Proposition **14**, we require from the above proof only the fact that each four rectangles of the family have a common transversal. With each line forming an angle ϕ with the two parallels, we associate a point on a circle having angular coordinate ϕ. The set of all ascending lines that intersect two given rectangles of the family is thus associated with a circular arc that is smaller than one third of a circle. Carried out for all pairs of rectangles from the family, this mapping produces a family of arcs that intersect pairwise because each four of the rectangles admit a common ascending transversal. There is a point common to all these arcs (Proposition **18**), and each two rectangles of the family admit a common ascending transversal parallel to the line L that corresponds to this point. Then under projection parallel to this line, the family of rectangles is carried onto a family of segments that have a common point by Proposition **16**. But then the line through this point parallel to L intersects all the rectangles of the family.

26. Let P be a point of a circle. To each line L in the plane, lay a parallel through P; let its second point of intersection with the circle be the image of the line L. Under this mapping, the set of all lines that intersect two fixed ovals goes into an arc. Carrying this out for all pairs of ovals from a family in which each four ovals have a common transversal, we obtain a family of pairwise intersecting arcs. There are two orthogonal directions corresponding to the antipodal pair of points that intersects all the arcs (Proposition **19**), and hence we conclude: *If each four ovals of a family admit a common transversal, then there exist two orthogonal directions such that each two ovals of the family admit a common transversal in one of these directions.* Now if the ovals of this family are mutually homothetic, and if four lines are laid out in the two orthogonal directions so that they form a rectangle circumscribed to a given oval of the family, then each of the family's (homothetic) ovals that is not smaller than the given one must be intersected by one of these four lines. Thus if there is a smallest oval of the family, the lines circumscribed to it meet all ovals of the family. If there is no smallest oval in the family, the desired result is obtained from some supplementary considerations on the limiting behavior of the size and position of the ovals in question. If the ovals are not only homothetic but are mutually congruent, it can be verified that some three of these four lines intersect all the ovals.

27. Let a line in the separating direction be chosen as the x axis. Every other line in the plane forms an angle ϕ (measured counterclockwise) with the x axis for which $0 \leqq \phi < \pi$. The set of all lines that intersect two ovals of the system, say A and B, corresponds on a ϕ axis to an interval of angles between 0 and π, which we denote by (AB), and similarly for other pairs of sets. We claim that each two of these intervals have a common point. Assuming this, there follows from Proposition **16** the existence of an angle ϕ_0 such that each two ovals of the system admit a common transversal in the direction ϕ_0. In other words, the parallel projections of the ovals in this direction form a system of pairwise intersecting segments on the x axis. Then all the ovals of the system are intersected by the projecting line through a point common to all the segments (Proposition **16**). It remains to show that each two intervals of angles have a common point. For such pairs of intervals as (AB) and (BC) this is assured by the assumption of a common transversal for A,B, and C. And if two intervals, say (AB) and (CD) should have no common point, then a contradiction would result as follows. Each of the intervals (AC), (AD), (BC), and (BD) would have points in common with both (AB) and (CD), so that the following situation arises for an angle ϕ' between (AB) and (CD): the ovals A and B, and also C and D, are separable by lines of direction ϕ', from which follows the separability of an additional pair by means of each of these two separating lines, but the pairs A and C, A and D, B and C, and B and D are not separable in this way. This is obviously a contradiction.

28. We prove the assertion first for a system of four ovals C_i ($i = 1,2,3,4$). Let L be a line that intersects C_1,C_2, and C_3. By M' and M'' we denote two lines that form an angle of $\pi/3$ with each other as well as with L. No line parallel to M' or M'' can intersect more than one of the ovals C_1,C_2 and C_3, since otherwise more than one of the ovals would subtend an angle of at least $\pi/3$ at the point where this line intersects L. The same argument shows also that either no parallel to M' or no parallel to M'' can intersect more than one of the ovals C_1,C_2,C_3,C_4; otherwise, a parallel L' to M' would intersect C_i and C_4 and a parallel L'' to M'' would intersect C_k and C_4, where i and k are among the numbers 1,2,3; a transversal M of C_i,C_k, and C_4, which must exist by hypothesis, then forms a nonobtuse angle $\geqq \pi/3$ with one of the lines L,L' and L''; M and this line would then both intersect the same two of the four ovals C_1,C_2,C_3, and C_4, which is impossible because of the condition on subtended angles. Thus the four ovals are totally separable either by parallels to M' or by parallels to M'', and according to Proposition **27** must admit a common transversal.

Let us prove the assertion for a system of more than four ovals. According to what has been proved already, we may assume that for each four ovals of the system there is a common transversal. Let P be a point of a circle. With each line L that intersects two ovals of the system we associate a parallel through P, and regard its second point of intersection with the circle as the image of the line L. In this way the set of all transversals common to two ovals is carried onto an arc; effecting this construction for all pairs of ovals of the system, we obtain a family of arcs, each smaller than one-third of a circle by the condition on subtended angles, and each two intersecting by the existence of a transversal for each four ovals. Thus all the arcs have a common point Q according to Proposition **18**, and the antipodal point Q^* is not in any of the arcs. Hence the line determined by the points P and Q^* yields a direction not corresponding to a transversal of any two ovals; therefore the system is totally separable by lines in this direction, and from Proposition **27** there follows the existence of a line that intersects all the ovals of the system.

29. Proposition **29** is a special case of Proposition **21**.

30. The lines can be replaced by sufficiently long segments; therefore, this is a special case of Proposition **21**.

31. In view of Proposition **29** it suffices to prove the assertion for a set of diameter 1 consisting of three points. If these form an obtuse triangle, its longest side is a diameter of a covering disk and hence $R \leqq \frac{1}{2}$. If the three points determine an acute triangle, then the circumcircle bounds a covering disk whose diameter is given by the formula $2R = a/\sin \alpha$, where a is the length of any side and α is the opposite angle. But every triangle has an angle $\alpha \geqq \pi/3$, whence $\sin \alpha \geqq \frac{1}{2}\sqrt{3}$ while $a \leqq 1$. Thus

$$2R = a/\sin \alpha \leqq 2/\sqrt{3}.$$

32. Proposition **32** is similarly reduced to the case of three lines with diameter 1. These form a triangle of perimeter $P \leqq 3$ that is circumscribed about the smallest intersecting circle. Since the equilateral triangle of perimeter $6r\sqrt{3}$ has the smallest perimeter of any triangle that can be circumscribed about a circle of radius r, it follows that $6r\sqrt{3} \leqq P \leqq 3$ and hence $r \leqq \frac{1}{6}\sqrt{3}$.

33. The point set may be assumed to be closed. Let S be an equilateral triangle that is circumscribed about the set, so that each of its sides includes a point of the set, and let S^* be another such triangle that is obtained by reflecting S in a point and then translating and magnifying, or contracting, if necessary, to obtain a second circumscribed equilateral triangle. Then either S or S^* has sides of length $s \leqq \sqrt{3}$. To see this, consider an arbitrary point that is common to S and S^*, and consider the perpendiculars from this point to the sides of the triangles. By a theorem from plane geometry, the sum of the three perpendiculars from any point in an equilateral triangle is equal to the altitude of that triangle. Since the set is of diameter $\leqq 1$, the sum of a perpendicular to S and the corresponding perpendicular to S^* must be $\leqq 1$, so that one of the triangles has altitude $\leqq 3/2$ and side of length $\leqq \sqrt{3}$.

34. Adding to the proof of Proposition **33**, we verify that the length of the side of the circumscribed equilateral triangle S varies continuously with the directions of the sides and becomes that of S^* after rotation through the angle π. Thus for some position of S, S and S^* are congruent and their intersection, which contains the given set of diameter 1, is a (possibly degenerate) centrally symmetric hexagon in which the distance between parallel sides is $\leqq 1$. This hexagon is contained in a regular hexagon that has the same center of symmetry and the same directions for its sides, and in which the distance between parallel sides is equal to 1. The regular hexagon has sides of length $1/\sqrt{3}$ and contains the given set.

35. Proposition **35** follows from Proposition **34**. In the regular hexagon of side length $1/\sqrt{3}$ that contains the set of diameter 1, construct three segments from the center that are perpendicular to the sides and form angles of $2\pi/3$ with each other. The hexagon is divided by these into three pentagons of diameter $\sqrt{3}/2$ that cover the given set.

The weaker form of Proposition **35**, corresponding to Borsuk's theorem for finite point sets, follows from Proposition **52** by mathematical induction. Since Proposition **52** asserts that among n points with $D = 1$ there must always be one which is at distance 1 from at most two others, then this one, once the other $n - 1$ points are already divided into three subsets of smaller diameter, can be added to the subset that does not include the two points; then all three sets are still of diameter < 1.

36. Let A be an oval that satisfies the hypotheses of the theorem and let S denote a strip of breadth $b < 1$ that covers A; we may assume that points P_1 and P_2 of A lie on the two lines that form the boundary of S. The connecting chord P_1P_2 divides A into two ovals, and neither of these can include two points Q_1 and Q_2 at distance 1. Otherwise the strip T of breadth 1 that covers Q_1 and Q_2 and is orthogonal to the chord Q_1Q_2 would also cover A; it would not be parallel to S, and A would be entirely contained in the parallelogram $S \cap T$. Further, since Q_1 and Q_2 lie on the same side of the line joining P_1 and P_2, the chords Q_1Q_2 and P_1P_2 would be two segments that join opposite sides of the parallelogram but do not intersect each other at inner points. From this it follows that at least one of the points P_1 and P_2 must be at a vertex of the parallelogram, contradicting the uniqueness of supporting lines.

37. If the oval A can be covered by parallel strips of breadth a and b, the assertion is evident. In the other case the crossing strips form a parallelogram having altitudes a and b; since A is convex, it is possible to pass a line through each vertex of the parallelogram in such a way that both A and the parallelogram are contained in the convex quadrilateral Q formed by these lines. Now each quadrilateral Q that is circumscribed about a parallelogram with altitudes a and b can be covered by a strip of breadth $c = a + b$. To see this we choose the notation so that the following conditions are satisfied (and such a choice is always possible): the vertex V_k of the parallelogram lies on the side G_k of the quadrilateral ($k = 1,2,3,4$); the parallelogram's opposite angles at V_2 and V_4 are not acute; of the two supporting strips S_2 and S_4 of Q that are parallel to G_2 and G_4 respectively, S_2 does not have greater breadth than S_4; the two end points of the segment G_1 are both on a boundary line of the supporting strip S_2. Now for each point of G_1, let us consider the sum of the distances from this point to the lines containing the segments G_2 and G_4; this sum, which at the end points of G_1 is equal to the breadths of S_2 and S_4 respectively, is a linear function on G_1, supposed to be coordinatized by means of a parameter representing length. Thus the breadth of S_2 is not greater than the sum of the distances of the point V_1 from the lines that contain G_2 and G_4. Moreover, this sum of distances, and even each individual distance, does not decrease if G_2 and G_4 are rotated about the vertices V_2 and V_4 respectively until both lines pass through the point V_3; in this position the sum of the two distances is equal to the sum of the altitudes of the parallelogram. Thus Q, and hence also A, is covered by the strip S_2 whose breadth does not exceed the value $c = a + b$.

38. Suppose the oval A cannot be covered by any strip of breadth 1, and let R be the radius of the largest circle that is contained in A. If this largest circle has two diametrically opposite points in common with the boundary of A, then the lines that support A at these points are tangent to the circle; the strip formed by these lines contains A and has breadth $b > 1$, whence surely $R \geqq 1/3$. Otherwise the circle and the boundary of A have in common three points that are not in any semicircle. Then the tangents to the circle at these three points form a triangle T that contains A and hence cannot be covered by any strip of breadth 1. Thus all the altitudes of the triangle are greater than 1 and the centroid of T, whose distance from any side is a third of the corresponding altitude, is the center of a circle of radius $1/3$ that lies entirely in T. Then of course for the inradius R of T, which is equal to the inradius of A, we have $R \geqq 1/3$. Thus in both cases A contains a circle of radius $r = 1/3$ about the incenter of A.

39. Let A be an oval and G an oriented line that makes an angle θ with a fixed reference direction. The breadth $\beta(A,\theta)$ of the oval A in the direction θ is the length of the normal projection of A on G. The value of this breadth averaged over all directions, that is

$$\beta(A) = \frac{1}{2\pi}\int_0^{2\pi} \beta(A,\theta)\, d\theta,$$

is called the average breadth of A. Computation for a segment S of length t shows easily that $\beta(S) = 2t/\pi$. From this it follows, first for polygons, and then by passage to the limit for arbitrary ovals, that the perimeter $L(A)$ and the average breadth $\beta(A)$ of an oval are related by the equation $\pi\beta(A) = L(A)$.

Now let $A_i (i = 1, \cdots, n)$ be a nonseparable system of ovals. If H is the convex hull of this system, then for each direction θ, the normal projection of the A_i's on a line G must cover the normal projection of H, since otherwise the system would be separated by some line orthogonal to G; consequently, the breadth $\beta(H,\theta)$ is less than or equal to the sum of the breadths $\beta(A_i,\theta)$. Forming the averages, we see that $\beta(H) \leqq \beta(A_1) + \cdots + \beta(A_n)$, and then from the relationship between β and L there results the inequality $L(H) \leqq L(A_1) + \cdots + L(A_n)$. This proves the theorem, for H covers the system of ovals.

40. Case 1. Suppose the circle contains an arc A of positive length that has no point in common with any arc of the family M. Note that the set of all end points of the arcs in M is at most countably infinite and that each neighborhood of a limit point of these end points must also include inner points of some of the arcs in M. Thus there exists an inner point P of A whose antipodal point P^* is an inner point of an arc B of positive length such that either B is a member of M or B is disjoint from all the arcs in M. Let Q^* and R^* be inner points of B that are on opposite sides of P^* and such that their antipodal points Q and R belong to A. Then the requirements of the theorem are satisfied by the two semicircular arcs that include P and are determined by the diameters QQ^* and RR^*, together with a semicircular arc that covers P^*, and also covers B in case B is a member of M.

Case 2. Suppose each point of the circle is a point of an arc in M or a limit point of such arcs, and suppose further that there is a point P such that P is an inner point of an arc A in M and the antipodal point P^* is an end point of an arc in M. Then we choose inner points Q and R of A on opposite sides of P_1, and the theorem's requirements are satisfied by the semicircular arcs that include P^* and are determined by QQ^* and RR^*, together with a semicircular arc that covers A.

Case 3. Suppose again that the circle contains no arc of positive length that is disjoint from all the arcs in M, but this time in addition that for each inner point P of an arc in M, the antipodal point P^* is an inner point of an arc in M. In this case the arcs in M form a system of antipodal pairs. If M includes two arcs A and B that can be covered by the interior of a semicircular arc C, we choose inner points Q and R of A and B; the theorem's requirements are satisfied by the semicircle C together with the two semicircles that are determined by Q and R, and their antipodal points Q^* and R^* whose intersection is the arc Q^*R^*. Finally, if no two arcs in M can be covered by the interior of a semicircular arc, then M must consist of four arcs that are antipodal in pairs and cover the entire circle. In this case the conditions of the theorem cannot be satisfied.

41. With each arc B of the family M we associate the polar arc B_0 of the same circle, where B_0 may be defined as the intersection of all semicircles whose mid-points (poles) lie in the arc B^* antipodal to B. Thus there arises a family M_0 of arcs that is easily seen to satisfy the hypotheses of Proposition **40.** Application of Proposition **40** to the family M_0 produces three semicircles, and the three points antipodal to the poles of these semicircles have the desired property with respect to the given family M of arcs. The exceptional cases described in Propositions **40** and **41** are the polar correspondents of each other.

42. We consider first the vertices of a proper oval A, these being the boundary points S at which A admits more than one supporting line. With each such boundary point we associate the smallest angle that contains A and is bounded by two rays emanating from S. Then we translate all the rays in this angle that emanate from S, so that they emanate instead from the center of a certain circle K of unit radius; their intersections with K form a closed arc that is smaller than a semicircle. The arcs associated with two different vertices S and T of A have a union that includes a pair of antipodal points, for S and T can be joined by a segment that lies in A and hence in the two associated angles. Except in the special case mentioned above, it follows from Proposition **41** that there are three points of K such that each arc in the family of arcs generated by the vertices of K has at least one of the points in its interior. We may characterize these points by means of their unit direction vectors $w_i (i = 1,2,3)$.

Let P be an arbitrary point of A and let $\phi_i = \phi_i(P)$ be the length of the "half chord" in which the interior of A is intersected by the ray that emanates from P in the direction w_i. If the ray intersects no interior point, we set $\phi_i = 0$. However, for each point P at least one of the three numbers ϕ_i is positive. If P is a vertex, then by the choice of the three directions w_i at least one of the three rays intersects the interior of A. If P is a regular boundary point, this follows from the fact that according to Proposition **41** not all of the three chosen directions can lie in a closed half space. And of course the assertion is obvious if P is an interior point of A. Now we define

$$\phi(P) = \max [\phi_i(P): i = 1,2,3],$$

so that ϕ is a lower semicontinuous positive function defined on A. Because A is compact there is a number $\sigma > 0$ such that $\phi(P) \geqq 2\sigma$ for all $P \in A$. Then the following can be said: If p is the position vector of a point $P \in A$, then at least one of the three points $p + \sigma w_i (i = 1,2,3)$ is always an interior point of A. This may be expressed alternatively as follows: If A_i is the oval that arises from A through translation by the vector $-\sigma w_i$, then each point $P \epsilon A$ is an interior point of at least one of the three ovals A_i. Thus our theorem is proved save for the exceptional case, and this arises exactly when A is a parallelogram.

43. From a proper oval A we can construct a centrally symmetric oval B by means of the Minkowski addition formula $B = \frac{1}{2}A + \frac{1}{2}\bar{A}$, where $\bar{A}$ is the oval that is symmetric to A with respect to a fixed origin; B can also be characterized as the set of all points with the position vectors $b = \frac{1}{2}(a_1 - a_2)$, where a_1 and a_2 run independently through all the position vectors of points of A. Two ovals A and A_w, where A_w arises from A through translation by the vector w, have common points or common interior points if and only if the same is true of the centrally symmetric oval B and the oval B_w that arises from it through translation by the vector w. In fact, from $a - a' = w$ with $a,a' \epsilon A$, which is equivalent to saying that A and A_w have the position vector a in common, it follows that $b - b' = w$ for $b = \frac{1}{2}(a - a')$ and $b' = \frac{1}{2}(a' - a)$, where $b,b' \epsilon B$. Conversely, from $b - b' = w$ where $b = \frac{1}{2}(a_1 - a_2)$ and $b' = \frac{1}{2}(a_3 - a_4)$ with $a_i \epsilon A$, it follows that $a - a' = w$ with $a = \frac{1}{2}(a_1 + a_4)$ and $a' = \frac{1}{2}(a_2 + a_3)$, where $a,a' \epsilon A$ by the convexity of A. The common point is an interior point provided the translation vector w can be chosen from a set of vectors whose end points, with respect to a fixed origin, fill some disk. Thus with respect to both intersecting and overlapping, the ovals A and A_w have the same relationship as the ovals B and B_w; it, therefore, suffices to prove the theorem for centrally symmetric proper ovals.

In order to show that $n(A) \leqq 9$, we choose as origin the center of a centrally symmetric proper oval A, and suppose that A is intersected by m nonoverlapping translates of A. All of these ovals lie in the oval $3A$ that is obtained from A by a dilation with the factor 3. In fact, the position vector of a point p of such an oval A_w can be expressed in the form $p = a + (b - c)$, where a is a point at which A_w intersects A and $(b + w) - (c + w)$ is a vector joining two points of A_w. This point belongs to $3A$ because it is the centroid of the triangles with vertices $3a, 3b$, and $-3c$, where a,b, and c are vectors of A, as is also $-c$ because of A's central symmetry. Since the m ovals do not overlap, we see that

$$mf(A) \leqq f(3A) = 9f(A),$$

where f represents area, whence $m \leqq 9$ and thus $n(A) \leqq 9$.

In order to show that $7 \leqq n(A)$ when A is a centrally symmetric proper oval, we circumscribe a centrally symmetric hexagon about A in such a way that the mid-point of each side of the hexagon is a point of A. This is possible, since for example the property is possessed by the centrally symmetric hexagon of smallest area that is circumscribed about A. At the six sides of this hexagon there can be adjoined six translation-equivalent hexagons that do not overlap and in fact belong to a lattice covering of the entire plane with such hexagons (see Figure 9). Each one of the six translates of A that are inscribed in these six hexagons has a mid-point of a side in common with A. Together with A itself, they constitute 7 translates of A that do not overlap although they all intersect A. From this it follows that $7 \leqq n(A)$.

We still want to show that $n(K) = 7$ for a circular disk K. The regular arrangement of six circles, all congruent and tangent to a seventh, shows that surely $n(K) \geqq 7$, corresponding to the general proof. Now consider the case in which m nonoverlapping congruent circles all intersect an additional one K. Among the m circles there are at least $m - 1$ whose centers $p_i (i = 1, \cdots, m - 1)$ do not coincide with the center of K. Considering a triangle having the centers p, p_i, and $p_k (i \neq k)$ as its vertices, we see that

$$|p - p_i| \leqq 2, \; |p - p_k| \leqq 2,$$

and $|p_i - p_k| \geqq 2$. Thus if ϕ_{ik} denotes the angle of the triangle at p, we have $\phi_{ik} \geqq \pi/3$. From this it can be derived that $m - 1 \leqq 6$ and thus $m \leqq 7$. Accordingly, $n(K) \leqq 7$, which together with the inequality $n(K) \geqq 7$, yields $n(K) = 7$.

44. A point set of diameter $D \leqq 2$ is formed by the centers of the disks of radius $R = 1$ that intersect pairwise. It follows from Proposition **34** that this set can be covered by a regular hexagon having sides of length $2/\sqrt{3}$. In this hexagon there are three points, the mid-points of three diagonals, at a mutual distance of 1 such that all points of the hexagon, and in particular the centers of the given disks, are at distance $\leqq 1$ from at least one of these points. Accordingly, each of the given disks includes at least one of the three points.

45. We consider the intersections $C_i = C \cap A_i (i = 1, \cdots, n)$. By hypothesis they are closed and at least two of them are nonempty, for example, C_1 and C_2. Since C is connected but can be expressed as the union of the closed sets C_1 and $C_2 \cup \cdots \cup C_n$, these sets must have a common point. Hence at least one point of C_1 and A_1 belongs to C_i and A_i for $i \geqq 2$.

46. From the condition on diameters we require only two simple consequences concerning the nature of the covering. First, each of the three covering sets $A_i (i = 1,2,3)$ must include exactly one of the vertices of the covered triangle T; let V_i be the vertex covered by A_i. Second, the side S_i opposite V_i must not include any point of A_i and hence must be entirely covered by the other two sets.

Let n be a natural number and let us cover the triangle T in a regular way by $m = 4^n$ equilateral triangles $T_i (i = 1, \cdots, m)$ having sides of length 2^{-n}. We shall show that at least one of these triangles has the property that all three of the sets A_i participate in covering it.

We attach the index j to a vertex of a triangle T_i provided the vertex belongs to A_j and j is the smallest number for which this is true. Then denote by p the number of triangles T_i that have all three of the indices 1,2, and 3 attached to their vertices, and by q the number of T_i's that have only the indices 1,2,2 or 1,2,1. Let $t_i (i = 1, \cdots, m)$ denote the number of sides of the triangle T_i that join two vertices having the indices 1 and 2; then $t_1 + \cdots + t_m = p + 2q$. To count the same number in a different way, consider all the sides of the triangles $T_i (i = 1, \cdots, m)$ whose end points have the indices 1 and 2; let u denote the number of these segments that lie on the boundary of the original triangle T and v the number that are interior to T. Then $t_1 + \cdots t_m = u + 2v$ and comparison with the earlier equality shows that $p + 2q = u + 2v$. We show further that p is odd. In fact, the segments, among those being considered, that lie in the boundary of T can only be in S_3, and conversely no point of index 3 can lie in S_3, Since the two end points of S_3 have the indices 1 and 2, it follows easily (for example, by induction on the number of division points in S_3) that u must be odd; hence p is also odd and in particular $p \neq 0$. Thus there are triangles T_i having the indices 1,2, and 3; choose one of these arbitrarily and denote it by T^n.

As can easily be deduced from the Bolzano-Weierstrass limit theorem, there is a point P of T such that each disk of positive radius about P contains some triangle of the sequence $T^n (n = 1,2, \cdots)$. Since the three vertices of T^n belong to the three sets A_i and since the A_i's are closed, P must belong to all three sets.

47. Let the three vertices of the triangle be $V_i (i = 1,2,3)$ and let S_i be the side opposite V_i. If A is one of the covering sets, there must be at least one side of the triangle that includes no point of A; otherwise, the diameter of A is at least 1. Now we divide the covering sets into three classes $C_i (i = 1,2,3)$, putting A in the class K_j if j is the smallest number such that A includes no point of S_j. Let A_i denote the union of all the covering sets in the class K_i. The conditions derived at the start of the proof of Proposition **46** from the assumption on diameters are obviously satisfied here also. Thus it follows from that proof that there is a point of the triangle that belongs to all three of the sets A_i. It must belong to at least three of the covering sets.

48. If P and Q are two points of A from which the closed set A is star-shaped and if R is an arbitrary point of A, then clearly the entire triangular region PQR is contained in A. In particular, A contains the segment TR for each point T of the segment PQ, and this shows that A is star-shaped from T. The set B of all points from which A is star-shaped is therefore convex and as a subset of A, it is obviously bounded. Finally, it is also closed, for if P is a limit point of points $P' \epsilon B$ and R is an arbitrary point of A, then since A is closed and contains each of the segments $P'R$, it must also contain the limit segment PR. Therefore, B is an oval.

49. Let the polygon be given a positive orientation, each side of the polygon being directed so that the polygon's interior lies to its left. Let S be a side of the polygon and H the left half plane that is bounded by the line determined by S. Then we choose a square Q that has one side on this boundary line, contains S, and in addition is so large that it entirely contains the part $P \cap H$ of the polygon P that lies in the half plane H. In this fashion, a square Q is associated with each side S of the polygon, and our hypothesis guarantees that each three of these squares have a common point. By Proposition **14** there is a point Z that belongs to all of the squares Q. The point Z cannot be exterior to the polygon and the segment joining Z to an arbitrary point of the polygon cannot run into the exterior of the polygon for otherwise, as is easily seen, Z would have to lie outside at least one of the squares Q. Thus all sides of the polygon are visible from Z.

50. Let A be the set with the stated property and let Z be an arbitrary but henceforth fixed point of A. We divide A into two disjoint subsets A' and A'' where for A' we choose the set of all points P for which the segment joining Z to P lies entirely in $A(ZP \subset A)$, while A'' is the set of all points $P \in A$ for which $ZP \not\subset A$. The set A' is obviously closed, for if P is a limit point of a sequence $P_i(i = 1,2, \cdots)$ with $ZP_i \subset A$, then $ZP \subset A$ because A is closed. If we could show that A'' is also closed then we would have a separation of A into two disjoint closed sets A' and A'', and since A is connected, one of these sets would have to be empty. But $Z \in A'$, so A'' must be empty; that is, $ZP \subset A$ for each point $P \in A$, and since Z was arbitrary A must be an oval.

Now the set A'' is closed. Under the contrary assumption there would be a point P with $ZP \subset A$ that is the limit point of a sequence $P_i(i = 1,2, \cdots)$ with $P_i \in A$ but $ZP_i \not\subset A$. Omitting certain points P_i if necessary, we can assume that all the points P_i belong to a closed half plane H whose boundary is the line containing the segment ZP. Since $ZP_i \not\subset A$, we can associate with each point P_i a point Q_i of ZP_i for which $P_iQ_i \subset A$ and which has, among such points, minimum distance from Z. Obviously, Q_i, which lies in H, is a boundary point of A, and since $ZQ_i \not\subset A$, the point Q_i cannot lie in a neighborhood U of Z for which $A \cap U$ is convex. Thus the sequence of points Q_i has on ZP a limit point Q that is different from Z. Let V be a circular neighborhood of Q for which $A \cap V$ is convex, let X be the point at which ZQ intersects the boundary circle of V, and let Q_n be a point of the sequence $Q_i(i = 1,2, \cdots)$ that is in V. Then Q_n lies in the interior of the half plane H, and a proper triangle QXQ_n is formed that is entirely contained in $A \cap V$. In the sequence Q_i there is a point Q_m that is so close to Q that the segment ZQ_m includes an interior point R_m of the triangle QXQ_n. Because of the convexity of $A \cap V$ the segment Q_mR_m must be contained in A, which contradicts the choice of Q_m on ZP_m. Thus A'' must be closed.

51. Let the set A have the stated property.

Case 1. Suppose A is not connected. Then $A = B \cup C$ where B and C are disjoint nonempty closed subsets. If one chooses three points $P,Q \in B$ and $R \in C$, then the segments PR and QR cannot lie entirely in A, and thus the segment PQ is covered by the set $A = B \cup C$. By the covering lemma, Proposition **45**, PQ must already be covered by one of the sets B and C, and in fact by B, for $P \in B$. Thus for each two points P and Q of B, the entire segment PQ is contained in B, and consequently B is an oval. The same reasoning can be carried out for C, and then with $A = B \cup C$ we have a representation of A as the union of two ovals.

Case 2. Suppose A is connected and is locally convex at each point $P \in A$. Then A is an oval by Proposition **50**.

Case 3. Suppose A is connected, the set S of all points P of A at which A is not locally convex is nonempty, and the convex hull $\bar{S}$ of S has interior points. From the fact that in each neighborhood of a point $P \in S$ there are two points P' and P'' whose connecting segment is not entirely contained in A, it follows easily that S must be closed and thus $\bar{S}$ is a proper oval. It follows also that the segment joining the point $P \in S$ to an arbitrary point $Q \in A$ must be contained in A, since with P' and P'' as chosen earlier either QP' or QP'' must lie entirely in A, and QP is a limit of such segments. This shows also that S is a subset of the set T of all points of A from which A is star-shaped. Since T is an oval according to Proposition **48**, not only S but also $\bar{S}$ must lie entirely in T. Thus S is a subset of A and each point P of S is a boundary point not only of A but also of $\bar{S}$. We note also that the segment joining two points Q' and Q'' of S cannot include any other point of S, for since the segment $Q'Q''$ lies in T, the intersection with A of a sufficiently small circular neighborhood of a point between Q' and Q'' consists of a segment, a semicircular region, or a full circular disk, and thus local convexity does not fail at the intermediate point.

Now let us choose a fixed point Z in the interior of $\bar{S}$. If X is an arbitrary point of $A - \bar{S}$, the segment XZ cannot include any point of S, for the points of S are boundary points of A while every point of XZ except for X itself is interior to the convex hull of X and a neighborhood of Z that lies entirely in $\bar{S}$, and this convex hull is contained in A. Thus the unique point P at which XZ intersects the boundary of $\bar{S}$ does not belong to S; however by Proposition **10** there are two points P' and P'' of S such that P lies on the segment $P'P''$. In this way we associate with each point $X \in A - S$ a unique segment U of positive length in the boundary of $\bar{S}$. The family of segments of positive length is countable, since at most finitely many can be of length at least $1/n$ times the perimeter of $\bar{S}$. Now all of these boundary segments $U_n (n = 1, 2, \cdots)$ of $\bar{S}$ will be divided into three subclasses $C_i (i = 1,2,3)$ so that two adjoining segments are never in the same class. This is accomplished by means of the following inductive procedure: Let U_1 belong to C_1. Then when $n > 1$ and each of the segments $U_1, \cdots, U_{n-1}$ has already been assigned to one of the three classes, let U_n belong to the class C_j where j is the smallest of the numbers 1,2,3 for which U_n has no end point in common with any of the segments $U_i (i < n)$ that have been assigned to the class C_j. In terms of this division into classes we form the set $B_i (i = 1,2,3)$ of all points $X \in A - \bar{S}$ that are associated with a segment of the class C_i by the construction formulated above. We form also the set $\bar{B}_i$ as the convex hull of the union $\bar{S} \cup B_i$, and finally the set A_i as the closure of $\bar{B}_i$, that is, the set of all points P such that each neighborhood of P includes a point $P' \in \bar{B}_i$, where possibly $P' = P$. Now we shall prove that the ovals A_i precisely cover A, that is, $A = A_1 \cup A_2 \cup A_3$. Consider an arbitrary point $P \in A_i$, each neighborhood of P including a point $P' \in \bar{B}_i$. By Proposition **10**, P' lies in a triangle $Q_1Q_2Q_3$ in which each vertex Q_k is either a point of $\bar{S}$ or is associated with a segment of the class C_i. Now the connecting segment Q_1Q_2 lies entirely in A. If Q_1 or Q_2 belongs to $\bar{S}$, this follows from the fact that $S \subset T \subset A$, while if $Q_1,Q_2 \in A - \bar{S}$ one reasons as follows: If the segments Q_1Z and Q_2Z intersect the same boundary segment of $\bar{S}$, where Z is the fixed point chosen earlier, then Case 2 above applies to the intersection with A of the triangular region Q_1ZQ_2; this is therefore convex and in particular contains Q_1Q_2. If, on the other hand, the segments Q_1Z and Q_2Z intersect different boundary segments of $\bar{S}$, then the angle Q_1ZQ_2 includes at least two different points $R_1,R_2 \in S$, namely the end points of the intersected segments. Since A is star-shaped from Z,R_1, and R_2, the quadrilaterals $ZQ_1R_1Q_2$ and $ZQ_1R_2Q_2$ must lie entirely in A. The segment Q_1Q_2 must be covered by one of these quadrilaterals; otherwise, R_1 and R_2 would have to be interior to the triangle Q_1ZQ_2 and either R_2 would be interior to the first quadrilateral or R_1 would be interior to the second, an impossibility since R_1 and R_2 are boundary points. Thus Q_1Q_2 lies in A, and the same can be said of Q_2Q_3 and Q_3Q_1. Since $Z \in T$, A must contain the entire convex hull of the points Q_1,Q_2,Q_3, and Z, and hence in particular $P' \in A$. Then $P \in A$ for P is a limit point of points P'. Conversely, from $P \in A$ it follows that $P \in A_i$ for $i = 1$, 2 or 3. This is trivial if $P \in \bar{S}$. If $P \in A - \bar{S}$, then P is associated with a boundary segment U_n, this with a class C_i, and thus $P \in B_i \subset A_i$.

Case 4. Suppose A is connected, the set S of points at which A is not locally convex is nonempty, and there is an open circular disk D such that the set $D \cap A$ is nonempty and lies entirely in some line L. Since the segment connecting a point $P \in S$ to a point $Q \in D \cap A$ is contained entirely in A, P must lie on L. This implies that S includes at most two points and that $D \cap A$ includes a point P_0 that is not a member of S. Now let B' be the convex hull of all points of A that are not on L and let B denote the closure of B'. Obviously B is an oval. This oval is entirely contained in A. In fact, if $Q \in B$ then each neighborhood of Q includes a point Q' of B', and Q' belongs to a triangular region $Q_1Q_2Q_3$ with $Q_i \in A - (A \cap L)$; since the segments joining Q_1 and Q_2 to the point P_0 mentioned above can have only the point P_0 in common with $D \cap A$, then Q_1Q_2 must be contained in A. The same holds for the segments Q_2Q_3 and Q_3Q_1. Thus A contains the convex hull of the points Q_1,Q_2, and Q_3 together with any point $P \in S$, and in particular $Q' \in A$. As a limit point of points Q', the point Q is also a member of A. Since the set $C = A \cap L$ is an improper oval with $A = B \cup C$, we have an expression of A as the union of two ovals.

Case 5. Suppose A is connected, the convex hull $\bar{S}$ (of the set of all points at which A is not locally convex) is nonempty and is contained in a line H, and there is no disk D of the sort described in Case 4. Let A_1^n (respectively, A_2^n) denote the set of all points of A whose distance from H is not less than $1/n$ and that lie to the left (respectively, right) of H. If the set A_i^n is nonempty ($i = 1,2$; $n = 1,2, \cdots$) it satisfies the assumptions of Case 1 or Case 2 above. Thus A_i^n is either an oval or the union of two disjoint ovals, and we may write

$$A_1^n = B_1^n \cup B_3^n \qquad \text{and} \qquad A_2^n = B_2^n \cup B_4^n,$$

where the sets $B_i^n (i = 1,2,3,4)$ are four mutually disjoint ovals that may be empty. Let the notation be chosen so that $B_i^n \subset B_i^m$ when $n < m$, and let B_i denote the closure of the union of all of the sets $B_i^n (n = 1,2, \cdots)$. Obviously B_i is an oval and

$$A = B_1 \cup B_2 \cup B_3 \cup B_4.$$

Indeed, if a point $P \in A$ lies on the line H, then each neighborhood of P includes points $P' \in A - (A \cap H)$. Such a point P' belongs to one of the sets B_i^n and hence to B_i, and the latter holds also for P as a limit point of points P'. If a point P of A is not in H, it is subject to the reasoning just applied to P'. Conversely, if $P \in B_i$ then $P \in A$ by construction.

If at least one of the ovals B_i is empty, the assertion of the theorem is proved. If none is empty, then two of the ovals can be joined to form a single oval. To see this, note first that each point $P \in S$ belongs to all four of the ovals B_i. Indeed, P lies on H, and since A is star-shaped from P, P is a limit point of nearby points in each of the sets B_i. By construction the sets B_1 and B_3 can have at most one common point on H, and consequently, S consists of a single point P_0. According to the hypotheses for the present Case 5, it is possible to choose three points $P_i \in B_i (i = 1,2,3)$ so that P_0 does not lie on any of the three connecting segments. One of the segments P_1P_2 and P_2P_3 must be entirely covered by A, and from this it follows that two of the sets B_i have on H a common point Q_0 that is different from P_0. We lose no generality in assuming that $Q_0 \in B_1 \cap B_2$, and then it can be shown that $B_3 \cup B_4$ is an oval. In fact, if we choose two points $Q_i \in B_i (i = 3,4)$, then in arbitrary neighborhoods of these there are points $Q_i' \in B_i (i = 3,4)$ that are not on H. Of the three segments Q_0Q_3', Q_0Q_4', and $Q_3'Q_4'$, the first two cannot be covered by A for otherwise B_1 and B_3 and likewise B_2 and B_4 would have a common point other than P_0. Thus the segment $Q_3'Q_4'$ lies in A. The part of $Q_3'Q_4'$ that lies in one of the half planes determined by H is covered by $B_1 \cup B_3$; since B_1 and B_3 have only the point P_0 in common, this part of the segment must be covered by B_3 alone. In the same way we conclude that the rest of the segment is covered by B_4. Hence Q_3Q_4, as a limit segment of segments $Q_3'Q_4'$, is entirely contained in $B_3 \cup B_4$. This union is therefore an oval, and the assertion of the theorem is satisfied by the expression

$$A = B_1 \cup B_2 \cup (B_3 \cup B_4).$$

Repetition of the same reasoning shows in addition that $B_1 \cup B_2$ is also convex.

All cases have now been exhausted and the theorem is proved.

52. For sets of 1,2, or 3 points the assertion is trivial. Now suppose that $n > 3$, the assertion has already been proved for sets of $n - 1$ points, and the set $P_1, \cdots, P_n$ has diameter $D = 1$. For each two points P_i and P_k whose distance has the value 1, draw the connecting segment P_iP_k. If at most two segments emanate from each point P_i, then the total number of segments is $\leqq n$, as claimed. If there exists a point, say P_1, from which at least three segments emanate, say to P_i, P_j, and P_k, let P_j lie inside the acute angle $P_iP_1P_k$. Now if a point P_m is at distance 1 from P_j, then P_jP_m must intersect both P_1P_i and P_1P_k, for two nonintersecting segments of length 1 always include a pair of end points at distance >1. It follows that P_m coincides with P_1; that is, P_j has distance 1 only from P_1. Then omitting P_j causes the loss of only one segment, while according to the inductive hypothesis, the remaining $n - 1$ points determine at most $n - 1$ segments of length 1.

53. First we show that a distance $a \neq 2\pi/3$ need not be realized. Let k be a natural number. We divide the circle C in a uniform way into $3k$ arcs $A_i (i = 1, \cdots, 3k)$ of length $2\pi/3k$, numbered consecutively with respect to a particular orientation. Let $U_j (j = 0,1,2)$ be the union of the k closed arcs A_i whose index i leaves remainder j on division by 3. The sets U_0, U_1 and U_2 clearly cover the circle, and in each the distances in the interval $2\pi/3k < a < 4\pi/3k$ are not realized by any pair of points. For $k = 1,2, \cdots,$ these intervals cover all distances in the interval $0 < a \leqq \pi$ with the sole exception of $a = 2\pi/3$. It remains to be shown that this distance is always realized. Let the circle C be covered by three closed sets S_0, S_1 and S_2. If one of these sets covers the entire circle, it surely realizes the distance $2\pi/3$. Otherwise Proposition **45** implies that some point P of the circle belongs to two of the covering sets, say to S_1 and S_2. Let Q and R be the two points of C that have distance $2\pi/3$ from P. If Q and R are both members of S_0, then S_0 realizes the distance in question. And if one of the two points, say Q, does not belong to S_0, then Q must lie in one of the sets S_1 and S_2; since P lies in the same set, that set must realize the distance $2\pi/3$ between P and Q.

54. Let A and B be closed subsets of the circle C that cover C. If one of these sets covers all of C, the assertion is evident. In the other case Proposition **45** guarantees the existence of a point P that belongs to both A and B. Now we want to show that if there is a distance $b(0 < b \leqq \pi)$ that is not realized by the set A, then the set B realizes every distance a of the interval $0 < a \leqq \pi$. Following the positive orientation of C, we fix a point R at distance b from P and a point S at distance a from R. Since P is covered by A, R cannot also be in A, and thus R is covered by B. If S also belongs to B, then B realizes the distance a between R and S, and the proof is complete. If, on the other hand, S is in A, we consider the point Q that is at distance a from P with respect to the given orientation. Since $S \epsilon A$ and QS determines the distance b, Q cannot lie in A. Thus it is covered by B, and since P also belongs to B this set must realize the distance a from P to Q, agreeing with the statement of the theorem.

55. Along with the disk, the bounding circle is also covered. The result follows from Proposition **54** upon observing that the interval of angular distance $0 < a \leqq \pi$ corresponds to the interval $0 < d \leqq 1$ of Euclidean distance.

56. We identify the unit segment S with the interval $0 \leqq x \leqq 1$ of an x axis. Let S be covered by the closed sets A and B, and let the left end point 0 belong to A. If the adjoining third of the interval $0 \leqq x \leqq 1/3$ is entirely covered by A, then the result is evident. Otherwise according to Proposition **45** there is a point p with $0 \leqq p \leqq 1/3$ that belongs to both A and B. We will show that if some distance a of the interval $0 < a \leqq 1/3$ does not occur in the set A, then every distance b of the interval $0 < b \leqq 1/3$ is realized in B. Since $p \epsilon A$, $p + a$ must lie in B. If B includes the point $p + a + b$, which is still in S, then the two last-named points realize the distance b in B, as claimed. But if $p + a + b$ belongs to A, then $p + b$ must lie in B because the distance a was excluded from A. Therefore, the distance b is realized in B between the point $p + b$ and the point p. This completes the proof.

57. A more direct proof can easily be given, imitating those of Propositions **54** and **56**. However, if one wishes to obtain Proposition **57** as a corollary of **56**, as indicated in the text, he may proceed as follows. Consider, in the line, a sequence $J_n(n = 1,2, \cdots)$ of intervals of length n, and apply Proposition **56** to the covering of J_n by the two sets in question, concluding that one of the two sets always yields the interval of distances $0 < d \leqq n/3$. According to the "pigeon-hole principle," asserting here that if infinitely many objects are distributed among a finite number of holes, then some hole includes an infinite number of them, at least one of the two sets will realize the interval $0 < d \leqq n/3$ for infinitely many values of n, and this set has the stated property.

58. Suppose the chord S joins the points P and Q of the continuum A. The set A' of all points of A whose distance from P is at most t is closed and includes P. The set A'' of all points of A whose distance from P is at least t is closed and includes Q. Since A is covered by A' and A'', there is, by Proposition **45**, a point R that belongs to A' as well as to A''. The segment joining P and R is a chord of the required length t.

59. If the theorem were false, there would be a covering of the plane by three closed point sets $A_i (i = 1,2,3)$ such that A_i includes no pair of points at distance d_i. Suppose $d_1 \leqq d_2 \leqq d_3$. About a point $Z \in A_3$ as center, or about an arbitrary point Z in case A_3 is empty, let us draw a circle C of radius d_3. Obviously C is covered by the two closed point sets $A'_1 = C \cap A_1$ and $A'_2 = C \cap A_2$. According to Proposition **54** at least one of these sets has the property of including, for each angular distance b of the interval $0 < b \leqq \pi$, a pair of points that realizes this distance. Suppose A'_j has this property. Then A'_j, and hence also A_j, includes a pair of points at Euclidean distance d_j, a contradiction that completes the proof.

60. Let us imagine, in contradiction of the theorem, that A is a closed set that includes no pair of points at distance d, but nevertheless the entire plane can be covered by five sets $A_i (i = 1, \cdots, 5)$ that are congruent to A. Let T be an equilateral triangle whose sides are of length $6d$; T is covered by the five intersections $A'_i = A_i \cap T$. None of the sets A'_i includes pairs of points at distance d, and since A'_i is closed and bounded there is even an $\epsilon > 0$ such that none of the sets A'_i includes a pair of points whose distance d' satisfies the condition $d - 2\epsilon < d' < d + 2\epsilon$. The parallel set $B_i = (A'_i)_\epsilon$, the set of all points P whose distance from the set A'_i is not more than ϵ, is closed, and it follows from the triangle inequality that B_i includes no pair of points at distance d. The components of B_i, that is, the maximal connected subsets of B_i, have diameters which by Proposition **58** cannot exceed the value d. Since each component contains a circle of radius ϵ, and since each two components of B_i are disjoint, B_i must have only finitely many components. This holds for $i = 1,2,3,4,5$. The triangle T and in particular a concentric homothetic equilateral triangle D' of side $2d$ are therefore covered by finitely many closed sets of diameter less than d, namely by the components of the sets B_i. By Proposition **47** there is a point Z in D' that is covered by at least three of these components, and they must be components of different sets B_i, say B_1, B_2, and B_3. Let C be a circle of radius d and center Z. Since Z was chosen in D', C lies entirely in D; because of the restriction on the diameters of components, C must be entirely covered by the sets $C \cap B_4$ and $C \cap B_5$. By Proposition **54** at least one of these sets yields all distances between 0 and $2d$, hence in particular the distance d. This contradicts the fact that none of the sets B_i includes a pair of points at distance d.

61. Let the vertices P,Q, and R of the given triangle be opposite the sides p,q, and r. The same notation will be used for all triangles obtained from the original one by motions of the plane. We may assume that $p \leqq q \leqq r$, and then from the triangle inequality $p + q \geqq r$ it is immediate that $2q \geqq r$. Suppose the plane is covered by the closed sets A and B. If one of these sets already suffices to cover the plane, the assertion is evident. In the other case there is a point Z of the plane that belongs to both A and B. We consider the two circles C_q and C_r of radii q and r about Z as center.

Case 1. One of the circles includes a point Y that belongs to both A and B. Then if we move the triangle PQR so that P coincides with Z and one of the points Q and R coincides with Y, it is obvious that all three of the points P,Q, and R lie in A or all lie in B according to whether A or B includes the vertex that does not coincide with Y or Z.

Case 2. Neither of the circles C_q and C_r includes a point that is common to A and B. Then by Proposition **45** each of C_q and C_r is already covered by one of the sets A and B. If they are both covered by the same set then the vertices of the triangle PQR all belong to this set in each position for which P coincides with Z. If, on the other hand, they are not covered by the same set—say C_q by A and C_r by B—then we consider the annulus bounded by C_q and C_r. By Proposition **45** it includes a point X that belongs to both A and B. Since $2q \geqq r$, a circle of radius q about X intersects both C_q and C_r, whence both A and B must participate in covering this circle. Thus it includes a point W common to A and B. Now bringing the triangle PQR into the position for which P coincides with X and R with W, we see that all three vertices lie in A or all in B, according to whether Q is in A or in B.

62. Suppose A_0 is a bounded closed set and $A_1 = \phi(A_0)$ is a proper subset of A_0 that is obtained from A_0 by means of a congruence ϕ. Then there exists a point $P_0 \in A_0 - (A_0 \cap A_1)$, and we must produce a contradiction. If the set $A_n = \phi^n(A_0)$ is obtained from A_0 by means of n-fold iteration of ϕ, then obviously

$$A_{n+1} \subset A_n \qquad (n = 0,1, \cdots).$$

The uniquely determined circumcircle C_0 of A_0 is carried by ϕ onto the circumcircle C_1 of A_1, and since A_1 is a subset of A_0, the circles C_0 and C_1 are identical and ϕ does not move the center Z of C_0. Thus ϕ must be a rotation or a reflection. Moreover, ϕ cannot be a reflection, for if it were we would have $A_2 = A_0$ while by construction P_0 belongs to A_0 but not to A_2. Thus only rotations need be considered. If w is the angle of rotation, then w/π cannot be rational for otherwise A_m would be identical with A_0 for a suitably chosen natural number m, again an impossibility since A_m is a proper subset of A_0. So suppose w/π is irrational, and consider the points

$$P_n = \phi^n(P_0) \qquad (n = -1,0,1,2, \cdots)$$

that are pairwise distinct and all different from Z. Obviously $P_n \in A_1$ $(n = 1,2, \cdots)$. Of the N points $P_n (n = 1,2, \cdots ,N)$ that all lie on a circle about Z, some two, say P_q and P_r, must have an angular distance of at most $2\pi/N$, and one can show that for large N this happens with $q - r > 1$. Since rotation through the angle $-(r+1)w$ carries the points P_q and P_r onto the points P_{q-r-1} and P_{-1}, this implies that the point P_{q-r-1} of A_1 has angular distance $\leqq 2\pi/N$ from P_{-1}. Letting N increase, we see that there are points of A_1 arbitrarily close to P_{-1}, and since A_1 is closed it follows that $P_{-1} \in A_1$. But

$$P_0 = \phi(P_{-1}) \in \phi(A_1) \subset A_1;$$

this contradicts the assumption that $P_0 \notin A_1$.

63. In contradiction to the theorem, suppose there is a bounded set C that can be decomposed into two disjoint sets A and B such that $C = A \cup B$ and $A \cap B = 0$ while all three of the sets A,B, and C are mutually congruent. Then there are two plane congruences ϕ and ψ such that $A = \phi(C)$ and $B = \psi(C)$. Since C is bounded, there is a circumcircle K of C, uniquely determined in size and position, that at the same time must be the circumcircle of A and of B for these sets are contained in C and congruent to C. The congruences ϕ and ψ must leave invariant the center Z of the circumcircle and hence must be rotations or reflections. We consider first the case in which both ϕ and ψ are rotations, whence $\phi\psi(P) = \psi\phi(P)$ for each point P. Then if P is a point of C, $\psi(P)$ belongs to B and $\phi(P)$ to A, and these two points of C are transformed further into points $\phi\psi(P)$ of A and $\psi\phi(P)$ of B. Since $A \cap B = 0$, it is not possible that $\phi\psi(P) = \psi\phi(P)$, and this case cannot arise. Thus at least one of the mappings, say ϕ, must be a reflection. Each point $P \in B$ is carried onto a point $\phi(P)$ of A, and that point onto $\phi\phi(P) \in A$. But if ϕ is a reflection, then $\phi\phi(P) = P$ and this contradicts the fact that $A \cap B = 0$.

64. Let $\phi(A) = A$, where A is a set containing more than one point and ϕ is a distance-increasing biunique transformation of A onto A. Assuming A bounded, we shall derive a contradiction. At least one of the points of A, say P_0, cannot be fixed under the transformation. Now we consider the doubly infinite sequence of points

$$P_n = \phi^n(P_0) \qquad (n = 0,1,-1,2,-2, \cdots)$$

where ϕ^n is the nth iterate of ϕ and ϕ^{-n} is the transformation inverse to ϕ^n. For the set A^* of all these points P_n we have $A^* \subset A$, $\phi(A^*) = A^*$, and

$A^* < A^*$. No two points P_n with different indices can coincide, for if they did, then A^* would reduce to a finite set and there would be a greatest distance realized between two points of A^*. The corresponding distance in $\phi(A^*)$ would have to be still greater, contradicting the fact that $\phi(A^*) = A^*$. Thus the set A^* is infinite. It is obvious also that (a) $d[P_{n-k},P_{m-k}] < d[P_n,P_m]$ for $k = 1,2, \cdots$, for indeed $P_n = \phi^k(P_{n-k})$ and $P_m = \phi^k(P_{m-k})$. In particular, $d[P_{-2},P_{-1}] < d[P_{-1},P_0]$, so that the number

$$\Delta = \frac{d[P_{-1},P_0] - d[P_{-2},P_1]}{2}$$

is positive. Now if A is bounded the sequence $P_1,P_2,P_3, \cdots$ includes a pair of points P_n and P_m for which $d[P_n,P_m] < \Delta$ while $m > n > 0$. Applying (a) for $k = m$ and $k = m + 1$, we obtain $d[P_{n-m},P_0] < \Delta$ and $d[P_{n-m-1},P_1] < \Delta$. By the triangle inequality,

$$d[P_0,P_{-1}] \leqq d[P_0,P_{n-m}] + d[P_{n-m-1},P_{n-m}] + d[P_{n-m-1},P_{-1}]$$

and hence (b)

$$d[P_0,P_{-1}] < 2\Delta + d[P_{n-m-1},P_{n-m}].$$

If $m - n = 1$, this implies

$$d[P_0,P_{-1}] < 2\Delta + d[P_{-2},P_{-1}]$$

and thus, recalling the definition of Δ, the contradictory relation $2\Delta < 2\Delta$. If $m - n > 1$ then (b) yields the same contradictory inequality upon application of (a) with $k = m - n - 1$.

65. In a bounded closed set A, containing more than one point, there are two points P and Q whose distance is equal to the diameter of the set. By a distance-increasing transformation these would be carried into two points P' and Q' of B whose distance exceeds the value $d[P,Q] = D(A)$. The diameter of B is at least $d[P',Q']$ and thus the relation $D(A) = D(B)$ is impossible.

66. The result will first be established for improper ovals, that is, for segments. For this purpose we identify A with the interval $-1 \leqq x \leqq 1$ of the real line and suppose in contradiction to the theorem that $A = U \cup V$, $U \cap V = 0$, and $V = \phi(U)$ where ϕ is a congruence. The transformation ϕ cannot be reflection in a point Z for the center Z would have to belong to A and hence to U or V; Z is carried into itself by the reflections ϕ and ϕ^{-1}, so that Z would have to belong to both U and V, whereas the assumption was that $U \cap V = 0$. We consider the other possible case, where ϕ is a translation by the amount w, and we may assume that $w > 0$. Obviously neither U nor V can include a pair of points at distance w because one of these points would be carried onto the other by ϕ or ϕ^{-1}, contrary to the fact that $U \cap V = 0$. Further, the intervals $J(1 - w < x \leqq 1)$ and $J'(-1 \leqq x < -1 + w)$ must be contained in V and U respectively because the points of these intervals are carried outside the set A by the transformations ϕ and ϕ^{-1} respectively. Now consider the $2N + 1$ points P_n having coordinates $nw(n = 0, \pm 1, \pm 2, \cdots, \pm N)$, where N is chosen so that $Nw \leqq 1 < (N + 1)w$. Then $P_{-N} \in J' \subset U$ and $P_N \in J \subset V$; since the total number of points P_n is odd there must be two adjacent points in the same subset U or V. This contradicts the fact that neither U nor V includes a pair of points at distance w.

Let A be a proper oval and again suppose $A = U \cup V$, $U \cap V = 0$, and $V = \phi(U)$ where of course ϕ is a congruence mapping the plane onto itself. By the proof for improper ovals there cannot be any line L that has points in common with A and is carried onto itself by ϕ. Consequently ϕ must be a rotation about a point Z. As a fixed point of the transformations ϕ and ϕ^{-1}, Z does not belong to either U or V and hence also not to A. There is one and only one point P of A that has smallest distance from Z. If $P \in U$, then $\phi(P)$ and P are equidistant from Z and therefore $\phi(P)$ cannot belong to A. This contradicts the supposition whereby $\phi(P)$ should have belonged to V. If, on the other hand, $P \in V$, then a contradiction arises from the same reasoning applied to the rotation ϕ^{-1}.

67. For $k = 1$ the assertion is trivially satisfied by $N_1(n,p) = n$. Now suppose $k > 1$. Then over the domain consisting of all pairs (n,p) of natural numbers, the function N_k is defined by means of the initial condition that $N_k(n,p) = n$ in case either $p = 1$ or $n \leqq p$ and the recursive formula

$$N_k(n + 1,p) = 1 + N_k(kN_k(n,p),p - 1)$$

for $n \geqq p$. The numbers defined in this way have the required properties. For $p = 1$ we can set $A_i = K_i(i = 1, \cdots, k)$; for $n \leqq p$, j is chosen so that K_j is nonempty and then A_i is taken to be empty for $i \neq j$ while A_j is a p-pointed subset of A belonging to the class K_j. Thus the initial conditions are accounted for and the statement of the theorem is verified, when it is not vacuous, which can happen for $n < p$. It remains to be shown for $n \geqq p$ that the statement holds for the number pair $(n + 1,p)$ when it is already known for the pair (n,p) and all pairs $(n',p - 1)(n' = 1,2, \cdots)$. Let us set $m = kN_k(n,p)$ and let A be a point set having at least $1 + N_k(m,p - 1)$ points whose p-pointed subsets are divided into the classes K_i. By the inductive assumption it is possible to choose k disjoint subsets A'_i of $A - (P)$, having altogether at least $m = kN_k(n,p)$ points, such that all the $(p - 1)$-pointed subsets of A'_i belong to K'_i. One of these k subsets, say A', must include at least $m/k = N_k(n,p)$ points, and by the inductive hypothesis there are k new subsets $A''_1, \cdots, A''_k$ of A'_1, having at least n points in all, such that all p-pointed subsets of $A''_i(i = 1, \cdots, k)$ belong to the class K_i. Now with $A_1 = A''_i \cup (P)$ and $A_i = A''_i(i = 2, \cdots, k)$, these sets have altogether at least $n + 1$ points and all p-pointed subsets of A_i belong to the class $K_i(i = 1, \cdots, k)$. Thus the proof is complete by mathematical induction.

The given values of $N_k(n,2)$ are easily obtained by use of the recursive formula in conjunction with the initial conditions $N_k(n,1) = n$ and $N_k(2,2) = 2$.

68. For $p = 1$ the theorem follows from the pigeon-hole principle. Now suppose $p > 1$ and the theorem is known for the case of $(p - 1)$-pointed subsets. Let the p-pointed subsets of the infinite set A be divided into classes $K_1, \cdots, K_k$. We shall construct recursively a decreasing sequence $A_0 \supset A_1 \supset A_2 \cdots$ of infinite subsets of A and an infinite sequence $P_0, P_1, P_2 \cdots$ of points of A. Let $A_0 = A$. For $i \geq 0$ choose an arbitrary point P_i from A_i and then form the set A_{i+1} as follows: For $j = 1, \cdots, k$, let K'_j denote the class of all $(p - 1)$-pointed subsets T of $A_i - (P_i)$ such that the set $T \cup (P_i)$ belongs to the class K_j; by the inductive hypothesis there is an infinite subset of $A_i - (P_i)$ whose $(p - 1)$-tuples all belong to the same class (say to K'_r), and this infinite set is chosen as A_{i+1}. Then the class K_r includes each p-tuple consisting of P_i and an arbitrary $p - 1$ points of A_{i+1}, and we regard the class K_r as being associated with A_{i+1}. By the pigeon-hole principle some class K_j must be associated with infinitely many of the sets A_i, say with $A_{i1}, A_{i2}, \cdots$; the points $P_{i1}, P_{i2}, \cdots$ then form an infinite subset of A whose p-tuples all belong to K_j.

69. Let A be the set of nodes of the graph, suppose the order N is at least $N_k(km,2)$, using the numbers introduced in Proposition **67**, and let each edge of the graph be represented by its pair of end points. By Proposition **67** it is possible to find k subsets A_i of A having a total of km points such that all edges joining points of the set A_i belong to the same class. By the pigeon-hole principle some one of these sets must have at least m points, and this is the desired conclusion.

70. This result appears as a special case of Proposition **69** if the edges of the complete graph of order N are divided into two classes according to their membership in the two complementary graphs.

71. The assertion is trivial for $n = 3$. We show that its validity for $n + 1$ follows from that for n. Let the edges $P_rP_s (r \neq s)$ of the complete graph $(P_0, \cdots, P_n)$ be divided in the indicated manner into the classes $K_i (i = 1, \cdots, k)$.

Case 1. There is a node, say P_0, that belongs to more than two edges of the same class, say K_k. The edges of the remaining complete graph $(P_1, \cdots, P_n)$ are then distributed among the $k - 1$ classes $K_i (i = 1, \cdots, k - 1)$, for none of these edges can have nodes in common with more than two edges emanating from P_0. Thus $k - 1 \geqq n - 2$ by the inductive assumption, or $k \geqq n - 1$ as claimed.

Case 2. There is no point of the sort required in Case 1. Then no class can contain more than three edges, and if m denotes the number of edges in the entire graph, we have $m \leqq 3k$; since $m = n(n + 1)/2$, it follows that $k \geqq n(n + 1)/6$. A simple computation shows that for $n \geqq 3$ this implies $k \geqq n - 1$.

72. Let the nodes of the graph A be divided into two classes K_1 and K_2 and suppose no two points of the same class are joined. Then however we form a finite sequence of nodes of A in which successive points are joined, the points belong alternately to K_1 and K_2. If such a sequence has the additional property that the terminal point and the initial point belong to the same class, which is surely the case when the two are identical, then the sequence consists of an odd number of terms. Thus A is an even graph.

Suppose conversely that the finite graph A is even and let P be a node of A. A path from P is determined by each sequence of nodes starting from P in which successive points are joined by an edge. Let P' be a point of A that cannot be reached by any path from P, P'' a point that cannot be reached from either P or P', and so forth. This procedure ends after a finite number of steps, and thus one obtains a finite subset A' consisting of nodes no two of which are joined by a path. Now let K_1 (respectively K_2) be the set of all points of A that can be reached from A' by a path consisting of an even (respectively, odd) number of edges. From the way in which A' was constructed it follows that each point of A belongs to either K_1 or to K_2. Further, these two sets are disjoint, for if a point $Q \in A$ could be reached from $P \in A'$ by a path Z_1 consisting of an even number of edges and from $P' \in A'$ by a path Z_2 consisting of an odd number of edges, then one could go from P along Z_1 and then in reverse order along Z_2 to arrive at P' by means of a path consisting of an odd number of edges. But if $P \neq P'$, there is no connecting path while if $P = P'$ there are only those having an even number of edges because A is even. Thus K_1 and K_2 form a partition of A. If two different points Q and Q' of the same class were joined by an edge of A, then it could be deduced that two points P and P' of A are joinable by a path consisting of an odd number of edges, and this was just shown to be impossible. Thus the partition is of the required sort.

73. Suppose the even graph A satisfies the hypothesis and consists of a total of e edges. If $e = k$, the assertion is evident. Suppose $e > k$ and make the inductive assumption that the theorem holds for all graphs A' having e' edges, with $e' < e$. Let P_0 and Q_0 be two points of A that are joined by the edge S_0. The omission of S_0 generates an even graph A' that satisfies the hypothesis and for which $e' = e - 1 < e$. The edges of A' can be divided into k classes $K_i (i = 1, \cdots, k)$ in the desired manner. Since the degree of the points P_0 and Q_0 is at most $k - 1$, at least one of the classes is not represented among the edges of A' that are attached to P_0 or Q_0. If both points are omitted from the same class, we can add S_0 to this class and obtain an admissible division of the edges of A. In the other case one can assume that the class K_1 is represented by an edge at Q_0 but not at P_0, and K_2 is represented at P_0 but not at Q_0. Now we consider a path $Q_0Q_1Q_2 \cdots Q_q$ that starts from Q_0, has edges alternately in K_1 and K_2, and is carried as far as possible in this manner; thus $Q_{2j}Q_{2j+1} \in K_1$ while $Q_{2j+1}Q_{2j+2} \in K_2$. Since the class K_1 is represented at Q_0 only by Q_0Q_1 and the class K_2 is not represented there at all, this path cannot return to Q_0. And in view of the characteristic property of the division into classes K_i, the path cannot return to any other point Q_i. Finally, the path in question cannot reach the point P_0, for if Q_q were to coincide with P_0 then since the class K_1 is not represented at P_0, the number q would be even; $P_0Q_0Q_1 \cdots Q_q$ would be a path in A for which the initial point and the terminal point coincide although its number of edges is odd and the graph A was assumed to be even. Therefore, since P_0 cannot be reached by the path, the edges of A' can be reclassified so that the class K_1 is not represented at either of the points P_0 and Q_0; then adding the edge S_0 to the class K_1 yields an admissible division of the edges of A. The desired reclassification does not affect the edges of A' that do not appear in the path $Q_0 \cdots Q_q$, but the edges $Q_{2j}Q_{2j+1}$ are shifted over to the class K_2 and the edges $Q_{2j+1}Q_{2j+2}$ to the class K_1.

74. If one joins by an edge the points that are associated with each other under the mapping, there arises a graph that is even by the necessary and sufficient condition given in Proposition **72**, and whose points are all of degree k. By Proposition **73** the edges can be divided into k classes so that no two edges of the same class meet at any point. The edges that belong to one of the k classes associate the points in pairs so that at each point of A or B exactly one edge appears, and in this way a biunique correspondence is set up.

75. With the two divisions of the set M into classes A_i and $B_i (i = 1, \cdots, p)$ of q points each, we associate the even graph G having the points P_i and $Q_i (i = 1, \cdots, p)$ with P_i joined to Q_j by s_{ij} edges where s_{ij} is the number of points in the intersection $A_i \cap B_j$. The points of G are all of degree q, and thus by Proposition **73** the edges of G can be divided into q classes such that at no point do two edges of the same class appear. Thus each class is represented exactly once at each point P_i and hence consists of exactly p edges. The edges belonging to a particular class correspond to the p points of M that satisfy the given condition.

76. The existence of $N(k,p)$ is trivial for $p = 2$, since a very simple use of the pigeon-hole principle shows that among $k + 1$ points there must be two that are in the same class, when only k classes are involved; thus it suffices to set $N(k,2) = k + 1$. Now take $p \geqq 2$ and make the inductive assumption that the existence of $N(k,p)$ is established for all $k = 2,3, \cdots$, and that $N(k,p)$ exceeds k. We show that also $N(k,p + 1)$ exists for all k and exceeds k, thus completing the proof. First some preliminary steps. It is convenient to regard all the series of points that appear as subsets of the integer positions

$x = 1,2, \cdots$ on an x axis. Suppose the points x are divided in a fixed way into k classes $K_\sigma(\sigma = 1, \cdots, k)$. We define the function f, taking k values over the set of x's, by putting $f(x) = \sigma$ when x belongs to the class K_σ. By a block $C = C(u)$ of length s we understand a series of consecutive points $x = u + \lambda(\lambda = 0,1, \cdots, s-1)$. The distance a between two blocks $C(u)$ and $C(v)$ of the same length s is defined by setting $a = |v - u|$. Two such blocks will be called K-equivalent provided

$$f(u + \lambda) = f(v + \lambda) \qquad (\lambda = 0, \cdots, s-1).$$

Basically, k^s different, that is, pairwise not K-equivalent, blocks of length s are possible. Thus if one is concerned with m consecutive blocks $C(u + t)$ $(t = 0, \cdots, s-1)$ of length s, these are divided into k^s classes, and if

$$m \geqq N(k^s, p)$$

then the series of blocks includes p equidistant blocks $C(v + \nu a)(\nu = 0, \cdots, p-1)$ where consecutive blocks have the distance a. In the following, these considerations will be repeated several times.

Now for the main proof. We first construct an increasing sequence of $k + 2$ numbers $m_i(i = 0,1, \cdots, k+1)$ by setting

$$m_0 = 1$$

and

$$m_{i+1} = 3N(k^{m_i}, p)(i = 0, \cdots, k).$$

Let $C_{k+1,0}$ be an arbitrary block of length m_{k+1}. There are p equidistant and pairwise K-equivalent blocks

$$C_{k,\nu} = C(u_k + a_k)(\nu = 0, \cdots, p-1)$$

of length m_k whose initial points $u_k + \nu a_k$ lie in the first third of the block $C_{k+1,0}$. The block $C_{k,p}$ need not be K-equivalent to the others, but by construction it does lie entirely in the block $C_{k+1,0}$, for it begins at the latest in the second third of that block and its length m_k satisfies, by the inductive hypothesis, the inequality $m_k < k^{m_k} < N(k^{m_k}p)$. This reasoning is repeated with the block $C_{k,0}$ in place of the block $C_{k+1,0}$, and so on. For $i = k, k-1, \cdots, 0$, there result in this manner blocks $C_{i,0}$ of length m_i such that the p equidistant and pairwise K-equivalent blocks $C_{i,\nu} = C(u_i + \nu a_i)$ all lie in the block $C_{i+1,0}$ and such that the block $C_{i,p}$ is also contained in $C_{i+1,0}$. The block $C_{0,0}$ that comes last in the construction consists of a single point u_0. For a point $x \in C_{i,0}$ we have

$$(\alpha)\ f(x) = f(x + a_i) \qquad (\nu = 0, \cdots, p-1),$$

and further

$$x + \nu a_i \in C_{i+1,0} \qquad (\nu = 0, \cdots, p).$$

Let $w_0 = u_0$ and

$$w_i = u_0 + pa_0 + pa_1 + \cdots + pa_i \qquad (i = 1, \cdots, k),$$

and let

$$z_{ij} = a_{i+1} + \cdots + a_j \qquad (i < j).$$

From several repeated applications of condition (α), it follows that

$$(\beta)\ f(w_i) = f(w_i + \nu z_{ij}) \qquad (\nu = 0, \cdots, p-1).$$

Since f takes on only k values, there must be two equal values among the $k+1$ numbers $f(w_i)(i = 0, \cdots, k)$. Suppose $f(w_r) = f(w_s) = \rho$, where it may be assumed that $0 \leqq r < s \leqq k$. Setting

$$y = w_r + \nu z_{rs} \qquad (\nu = 0, \cdots, p),$$

we have $y_0 = w_r$ and $y_p = w_s$, and conclude easily in view of (β) that

$$f(y_\nu) = f(y_0) \qquad (\nu = 0, \cdots, p).$$

Thus a series of $p + 1$ equidistant points has been found such that all belong to the same class K_ρ. These have been found in an arbitrary block of length m_{k+1}, so it suffices to set $N(k,p+1) = m_{k+1}$.

77. Let the points of the unit lattice U be divided into k classes. Let a and b be two integers, let p,q,u and v be natural numbers and let $R(p,q;\, u,v)$ be the subset of U consisting of all points (x,y) that can be represented in the form

$$x = a + \nu u,\ y = b + \mu v \qquad (\nu = 0, \cdots, p-1, \mu = 0, \cdots, q-1).$$

Thus R is a rectangular section of a rectangular sublattice of U. Since a finite subset S of U is always a subset of a suitably chosen set $R(p,q;\, 1,1)$, and since an arbitrary set $R(p,q;\, u,v)$ is an affine image, it clearly suffices to show that for any two prescribed natural numbers p and q it is always possible to find an $R(p,q;\, u,v)$ whose points all belong to one and the same class. In order to show this we first choose $N = N(k,p)$ according to Proposition **76**. In the series of points $x = 1, \cdots, N$, $y = n$, where n at first is a fixed natural number, there are p equidistant points

$$x_\nu = a + \nu u \qquad (\nu = 0, \cdots, p-1)$$

that all belong to the same class. If we let n run through all the natural numbers, there results a sequence of equidistant series of points in which the coordinate $a = a(n)$ and the differences $u = u(n)$ of consecutive points, and also the indices $\sigma = \sigma(n)$ that indicate the membership of the points of the series in one of the k classes K_σ, are all functions of $n(n = 1,2, \cdots)$. Since $1 \leqq a \leqq N, 1 \leqq u \leqq N$, and $1 \leqq \sigma \leqq k$, there are kN^2 different triples of values, and accordingly the ordinates $y = n$ can be divided into kN^2 classes. In the interval $1 \leqq y \leqq N(kN^2,q)$ there are q equidistant numbers $y = b + \mu v$

($\mu = 0, \cdots, q - 1$) that belong to the same class. Let the triple of values corresponding to this class be a,u, and σ. By construction the points

$$x = a + \nu u, y = b + \mu\nu \qquad (\nu = 0, \cdots, p - 1, \mu = 0, \cdots, q - 1)$$

belong to the class K_σ and they form a set $R(p,q; u,v)$ of the desired sort.

78. We first introduce two terminological abbreviations that will be useful in later proofs also. We say that a family of sets has the (p,q) property if among each p sets in the family there are always q that have a nonempty intersection. The (p,q) property subsumes the $(p - r, q - r)$ property for $r = 1,2, \cdots, q - 1$, for starting from any choice of $p - r$ sets by which no point is covered at least $q - r$ times, the adjunction of an arbitrary r additional sets produces a choice of p sets by which no point is covered at least q times. Further, we say that a family of sets is n-partitionable provided the sets in the family can be divided into n (possibly empty) subfamilies so that the sets in each nonempty subfamily have at least one common point.

Now let S be a family of segments (in a line) that has the (p,q) property. If $p = q \geqq 2$, then S has also the (2,2) property and thus by Proposition **16** there is a point common to all the sets of the family. Hence the assertion is true for $p - q + 1 = 1$. Now consider a family S of segments having the (p,q) property for $p = p_0$ and $q = q_0(p_0 > q_0 \geqq 2)$, and make the inductive hypothesis that the theorem is already established for all $p \geqq q \geqq 2$ with $p - q < p_0 - q_0$. Suppose at first that S is finite. Let P denote the point of the line that is the farthest to the left among the right-hand end points of the members of S, and let S' denote the subfamily consisting of all members of S that include P while S'' consists of all members of S that do not include P. Then the family S'' is $(p_0 - q_0)$-partitionable. This is trivial if S'' includes at most $p_0 - q_0$ segments. And if S'' includes more than $p_0 - q_0$ segments then S'' has the $(p_0 - q_0 + 1, 2)$ property, for S'' lies entirely to the right of P, and if to an arbitrary choice of $p_0 - q_0 + 1$ segments of S'', there is added the segment that has P as right-hand end point and hence is disjoint from all of them, then some two of these segments must have a common point because S has the $(p_0 - r, q_0 - r)$ property for $r = q_0 - 2$. Thus by the inductive hypothesis S'' can be partitioned into

$$(p_0 - q_0 + 1) - 2 + 1 = p_0 - q_0$$

subfamilies of segments each having nonempty intersection. Together with the subfamily S' these form a $(p_0 - q_0 + 1)$-partition of S.

If S is not finite, let R denote the set of right-hand end points of the segments that are members of S. Certainly R is bounded on the left for otherwise infinitely many disjoint segments would appear in S. Thus there is a point P' that is farthest to the right among those points that have no point of R to the left of them. With P' it is possible to effect an argument similar to that involving P above, and then the proof is complete by mathematical induction.

79. If P is a point of the circle, all the arcs of the family that do not include P can be mapped by stereographic projection onto segments in the line that is tangent to the circle at the point antipodal to P. This family of segments admits a $(p - q + 1)$-partition, a fact that is trivial if it includes at most $p - q + 1$ segments. If it includes at least $p - q + 2$ segments, then note that like the family of arcs it has the (p,q) property and hence also the $(p - r,q - r)$ property for $r = q - 2$, whence by Proposition **78** a $(p - r) - (q - r) + 1 = (p - q + 1)$-partition is possible. Thus $p - q + 1$ points can be chosen so that all the segments in any one of the subfamilies have one of these points in common; the corresponding points on the circle together with P fulfill the requirements of the theorem.

80. It is to be proved that a family R of rectangles with the (p,q) property for $p \geqq q \geqq 2$ always admits an n-partition with

$$n = \frac{(p - q + 1)(p - q + 2)}{2}.$$

If $p = q \geqq 2$, then R has also the (2,2) property and the assertion becomes that of Proposition **15**. Now suppose that $p_0 > q_0 \geqq 2$ and that the theorem has already been proved for all $p \geqq q \geqq 2$ with $p - q < p_0 - q_0$. Let R be a finite set of mutually parallel rectangles having the (p_0,q_0) property. There is a half plane H such that H is parallel to the sides of the rectangles and H covers a rectangle A from the family R although no member of R lies entirely interior to H. Let R' be the subfamily consisting of all rectangles of R that intersect H, and let R'' be the remaining family of rectangles of R that do not intersect H. The orthogonal projections of the members of R' onto the bounding line L of H are identical with the segments in which these rectangles intersect L. If the family S' of these segments has at least $p_0 - q_0 + 2$ members, then S', just as R, has the $(p_0 - r,q_0 - r)$ property for $r = q_0 - 2$, so by Proposition **78** S' and hence R' admits a $(p_0 - q_0 + 1)$-partition. If S' includes fewer than $p_0 - q_0 + 2$ segments then a $(p_0 - q_0 + 1)$-partition of R' is trivial. It can be proved also that R'' is $[(p_0 - q_0)(p_0 - q_0 + 1)/2]$-partitionable. If R'' includes at most $p_0 - q_0$ rectangles, this is trivial for $p_0 - q_0 + 1 \geqq 2$. And if R'' includes more rectangles, then R'' has the $(p_0 - q_0 + 1,2)$ property for each one of the $p_0 - q_0 + 1$ rectangles of R'' together with the rectangle A, which is covered by H and disjoint from them, provides a choice of $p_0 - q_0 + 2$ rectangles of R, and R has the $(p_0 - r,q_0 - r)$ property for $r = q_0 - 2$. Thus by the inductive hypothesis, R'' can be divided into

$$\frac{[(p_0 - q_0 + 1) - 2 + 1][(p_0 - q_0 + 1) - 2 + 2]}{2} = \frac{(p_0 - q_0)(p_0 - q_0 + 1)}{2}$$

classes of rectangles each having a common point. Together with the $p_0 - q_0 + 1$ classes of R' these classes form a $[(p_0 - q_0 + 1)(p_0 - q_0 + 2)/2]$-partition of R. For an infinite family of rectangles one encounters limit considerations in choosing the half plane H but otherwise the proof is the same.

81. We will call a number pair p,q admissible provided $2 \leqq q \leqq p \leqq 2q - 2$. Obviously the admissibility of p_0,q_0 implies that of $p_0 - r, q_0 - r$ for $0 \leqq r \leqq 2q_0 - 2 - p_0$. Now suppose the family R of parallel rectangles has the (p,q) property for an admissible pair p_0,q_0. If $p_0 = q_0$, the assertion of the theorem follows from Proposition **15**. Suppose $p_0 > q_0$. Then with $r = 2q_0 - 2 - p_0$ and $j = p_0 - q_0 + 1 > 1$, we have $p_0 - r = 2j$ and $q_0 - r = j + 1$, so R has the $(2j, j + 1)$ property. Proceeding by induction, we assume that the theorem has already been proved for admissible pairs $p = 2i, q = i + 1$ when $i < j$, which surely is the case for $j = 2$.

Case 1. Suppose R has not only the (p_0,q_0) property but also the $(2(j - 1), j)$ property. Since the latter number pair is admissible, by the inductive hypothesis R admits a

$$p - q + 1 = 2(j - 1) - j + 1 = (p_0 - q_0)\text{-partition},$$

which is the desired conclusion.

Case 2. Suppose R lacks the $(2(j - 1), j)$ property, so that there is a subfamily R' of R consisting of $2j - 2$ rectangles $A_1, \cdots, A_{2j-2}$ by which no point is covered as many as j times. Considering R' together with two arbitrary rectangles from $R - R'$ and recalling that R has the $(2j, 2j + 1)$ property, we observe first that each two rectangles from $R - R'$ have a common point and secondly that some $j - 1$ rectangles from R', say $A_j, \cdots, A_{2j-2}$, have a common point. A j-partition of R is then obtained as follows. For $n = 1, \cdots, j - 1$, let the subfamily R_n include A_n and all the rectangles of $R - R'$ that have a point in common with A_n and have not already been allocated to one of the systems $R_1, \cdots, R_{n-1}$. Let R_j include $A_j, \cdots, A_{2j-2}$ and all rectangles of $R - R'$ that are disjoint from $A_1, A_2, \cdots$, and A_{j-1}. The rectangles of each of these subfamilies have a common point. For $n = 1, \cdots, j - 1$ this follows from Proposition **15** in conjunction with the first observation above and the method of construction of R_n. For R_j one reasons as follows. If R_j includes only $A_j, \cdots, A_{2j-2}$, apply the second observation above. If R_j includes in addition at least one rectangle A from $R - R'$, then A together with R' and an arbitrary additional rectangle from $R - R'$ form a selection of $2j$ rectangles, and these must cover some point at least $j + 1$ times. If this point was not in A, then it would be covered at least j times by R', contradicting the definition of R'. Thus it is included in A, and since A is disjoint from $A_1, \cdots, A_{j-1}$, it must be covered by $A_j, \cdots, A_{2j-2}$ and the additional rectangle from $R - R'$. Thus A has at least one point in common with $A_j, \cdots, A_{2j-2}$ and each additional rectangle from $R - R'$, in particular, each additional rectangle from

R_j. By Proposition **15** there is a point that belongs to all the rectangles from R_j, and this completes the proof.

82. If the family includes fewer than n rectangles, the assertion is trivial. For n or more rectangles this result is the special case $p = n + 1$, $q = 2$ of Proposition **80**.

83. We prove the theorem by complete induction on the number k of rectangles that belong to the family. The conclusion is obvious for $k \leqq m + 1$. Now suppose $k > m + 1$, A is a rectangle from a family R of k parallel rectangles satisfying the given hypotheses, and the theorem has already been proved for families of fewer than k rectangles. In particular, the theorem applies to $R - (A)$ and $R - (A)$ can be divided into $m + 1$ subfamilies so that the rectangles of each subfamily have a nonempty intersection. The theorem is proved by showing that A can be allocated to one of these subfamilies without loss of the intersection property. If A is disjoint from the nonempty intersection of such a subfamily, then by Proposition **15** it is also disjoint from some rectangle in the family. Accordingly, A cannot be disjoint from the intersection of each of the $m + 1$ subfamilies, for the family does not include $m + 1$ rectangles that are disjoint from A.

84. This is proved by means of the more special result **85**.

85. Here we call a number pair p,q admissible in case $3 \leqq q \leqq p \leqq 2q - 3$. Obviously the admissibility of p_0,q_0 implies that of $p_0 - r, q_0 - r$ when $0 \leqq r \leqq 2q - 3 - p$.

Now suppose the family E of ovals has the (p_0,q_0) property for an admissible pair (p_0,q_0) (compare the proof of Proposition **78**). If $p_0 = q_0$, the desired conclusion holds by Proposition **14** with $p_0 - q_0 + 1 = 1$. Suppose $p_0 > q_0$. With $r = 2q_0 - 3 - p_0$ and $j = p_0 - q_0 + 1 > 1$, E has also the $(2j + 1, j + 2)$ property for $p_0 - r = 2j + 1$ and $q_0 - r = j + 2$. We must show that there is a j-partition of E, and we assume inductively that the theorem is already proved for all admissible pairs $p = 2i + 1$, $q = i + 2$ with $i < j$. Surely this holds for $j = 2$.

For the given family E of ovals $A_k (k = 1, \cdots, n)$ we now construct an oval U that satisfies the following conditions: (a) U has a point in common with each $A_k \in E$; (b) the family E_U of ovals $A_k \cap U (k = 1, \cdots, n)$ has the $(2j + 1, j + 2)$ property; (c) U is an oval of minimum area among those with the properties (a) and (b). By choosing a sufficiently large square, one is easily convinced of the existence of an oval U that satisfies the first two conditions. Because of the closedness and the finite number of the ovals A_k it is also possible to satisfy all three conditions. The oval U is a proper convex polygon or a segment. In fact, if one chooses a point in each set $A_k \cap U$ $(k = 1, \cdots, n)$ and, further, for each choice of $2j + 1$ ovals from E_U, chooses a point belonging to at least $j + 2$ of these, then the convex hull of these points is a polygon that lies entirely in U and satisfies the conditions (a) and (b). Because of (c) this polygon must be identical with U or with a segment. In the last case the ovals $A_k \cap U$ form a family of segments with the $(2j + 1, j + 2)$ property, admitting a j-partition by Proposition **78**; a fortiori, the family E has the same property. Thus we may assume henceforth that U is a proper polygon. Let P denote a vertex of U and let E'_U denote the collection of all ovals of E_U that include P while E''_U denotes the remaining collection of ovals of E_U. Now we show that either E'_U consists of a single oval that is identical with P (first case) or that there are two ovals in E'_U that have no common point other than P (second case). Indeed, the vertex P must belong to at least one oval from E_U because of the minimum condition (c). If it belongs to only one, then this one must be identical with P, again by the minimum condition (c). On the other hand, if P belonged to several ovals from E_U and each two of these ovals had a common point Q_μ different from P, then all of the connecting segments PQ_μ could be intersected by a line L missing P. From each oval of E'_U the line L would cut a segment, or a point, and each two of these segments would have a common point, namely the intersection of L with PQ_μ. Then by Proposition **14** there would be a point Q of L common to all the segments so that all the ovals of E'_U would have a common point other than P. But in this case U would not be minimal. Thus, as claimed, E'_U must include two ovals that have only the point P in common.

Case 1. E'_U consists only of the oval P. Choosing, for each selection of $2(j - 1) + 1$ ovals from E''_U, an additional oval of E''_U and also the point P, one obtains a selection of $2j + 1$ ovals from E_U; by (b) there are $j + 2$ of these that have a common point, and since this is different from P, it must be in at least $j + 1$ ovals from the original selection of $2(j - 1) + 1$. Hence E''_U has the $(2i + 1, i + 2)$ property for $i = j - 1$ and thus by inductive hypothesis it admits a $(j - 1)$-partition. Together with the class $E'_U = (P)$ this forms a j-partition of E_U, and by (a) this is also the desired j-partition of E.

Case 2. E'_U includes two ovals A' and A'' that have only P in common. In order to show the possibility of a j-partition of E it suffices as earlier to show that E''_U admits a $(j - 1)$-partition. If E''_U includes at most $j - 1$ ovals, this is trivial. If E''_U consists of $2j - s$ ovals $(2 \leqq s \leqq j)$, one adds to E''_U the ovals A' and A'' as well as an additional $s - 1$ ovals from E'_U, obtaining thereby a selection of $2j + 1$ ovals from E_U. By (b) some $j + 2$ of these ovals have a common point, and this must be different from P for P lies in only $j + 1$ ovals among those selected. The point, covered at least $j + 2$ times, must belong to at least $j + 2 - s$ ovals from E''_U. The required $(j - 1)$-partition of E''_U is obtained by taking the family of these $j + 2 - s$ ovals together with the remaining $j - 2$ ovals from E''_U, each of the latter being regarded as a family of ovals having but a single member. Now suppose finally that E''_U includes at least $2j - 1$ ovals. Then E''_U has the $(2i + 1, i + 2)$ property for

$i = j - 1$ and thus admits a $(i - 1)$-partition by the inductive hypothesis. Indeed, to $2j - 1$ ovals from E''_U one adds the two ovals A' and A'' of E'_U that have only the point P in common. By (b) this selection of $2j + 1$ ovals from E_U includes $j + 2$ that have a common point. This point cannot coincide with P. Hence it must lie in only one of the ovals A' and A'' and consequently belongs to at least $j + 1$ of the ovals chosen from E''_U. This is the desired conclusion.

86. Let Z denote the center of the circle, and let A denote the convex hull of the closure A of A.

Case 1. Z lies on the boundary or in the complement of the convex domain A. Then A lies in a closed half plane that is bounded by a line through Z. From this it follows immediately that A lies in a closed semicircular arc.

Case 2. Z is an interior point of A. By Proposition **11** Z is an interior point of the convex hull of three or four points of A. However, Z cannot be an interior point of a triangle formed by three points of A, for then there would be three points of A that are not in any semicircle, contradicting the hypotheses. Thus Z must be an interior point of a quadrilateral formed by four points of A, and indeed by the same reasoning Z must coincide with the point of intersection of the two diagonals. Consequently A includes two pairs of antipodal points P,P^* and Q,Q^*. But then A consists of only these four points, for if there were an additional point R of A, it could be assumed without loss of generality that R is interior to the shorter circular arc joining P^* to Q^*, and the triangle having vertices P, Q, and R would then have Z in its interior, an impossibility. Thus A consists of the four points P,P^*,Q,Q^*, and the results corresponding to the two cases cover the two possibilities claimed in the statement of the theorem.

87. The statement is correct for parallelograms, for then the family of ovals turns out to be a family of mutually parallel parallelograms, and for such families the conclusion follows from Proposition **15** with the aid of an affine transformation. If A is not a parallelogram, we must show that the conclusion fails for certain families that satisfy the hypotheses. For this it suffices to exhibit three translates A_i of $A (i = 1,2,3)$ that have an empty intersection but do intersect pairwise. Let Z be an interior point of an oval A that is not a parallelogram and let C be a circle with Z as center. If P is a regular boundary point of A, then A admits a unique supporting line L at P. Let Q be the point at which the circle C is intersected by the ray from Z perpendicular to L. The points Q corresponding to the various regular boundary points P of A form a subset D of C that for a proper oval A cannot lie entirely in a closed semicircular arc of C. Thus it is possible to find three points Q_i of $D (i = 1,2,3)$ that do not belong to any semicircle, for otherwise by Proposition **86** D would have to consist of two pairs of antipodal points, and from this it could be deduced that A must be a parallelogram. If the points $P_i(i = 1,2,3)$ are the three boundary points of A that are in biunique correspondence with the points Q_i, then we translate A to the three positions $A'_i = \phi_i(A)(i = 1,2,3)$, where ϕ_i denotes that translation of the plane for which $\phi_i(P_i) = Z$. Each two of the ovals A'_i have a common interior point but Z is the only point common to all three. If we then move the domains A'_i by sufficiently small amounts in the directions opposite to ZQ_i, the intersections of each two of the three will still be nonempty while the intersection of all three will be empty.

88. Each two ovals from the family must have nonempty intersection, for otherwise the hypothesis would not be satisfied by the parallels to a separating line. When A is a parallelogram, the conclusion follows from this observation in conjunction with the affine variant of Proposition **15**. If A is not a parallelogram, then by the proof of Proposition **87** there are three translates A_i of $A(i = 1,2,3)$ that have a pairwise nonempty intersection although no point is common to all three. Consider an arbitrary line L. By orthogonal projection of the three nonempty intersections $A_1 \cap A_2$, $A_2 \cap A_3$, and $A_3 \cap A_1$ onto a line T perpendicular to L, we obtain three segments T_i of T. If two of these segments have a common point P', then the line through P' parallel to L intersects all three of the ovals A_i. If the three segments in T are pairwise disjoint, we choose P' in the middle segment and reason analogously. Thus the assumption about parallel lines is satisfied even though the three ovals have empty intersection, and this completes the proof.

89. The "only if" part is easily verified. If there is a line that intersects all the ovals of the family, then the disjoint linearly ordered segments or points of intersection determine an order among the ovals, and each three ovals are intersected in the given order by the line in question.

In proving the "if" part, we will say that a family E of disjoint ovals A has the transversal property provided it can be ordered in such a way that each three ovals of the family are intersected in the given order by some line. Obviously the transversal property carries over from E to each family F of disjoint ovals such that each member $B \in F$ contains a member $A \in E$; one need only transfer the order relation from E to F in the natural way.

Now in each oval $A \in E$ we choose a point and then replace A by the oval λA obtained from A by a contraction toward this point in the ratio $\lambda : 1 (0 \leqq \lambda \leqq 1)$. The ordering in E can be transferred in a natural way to the family $E(\lambda)$ of contracted ovals $\lambda A (A \in E)$, where each oval A is contracted toward its own individual point. First let us suppose that E is a finite family and let the numbering $A_1, A_2, \cdots$ of the ovals of E correspond to the order relation. A simple convergence argument shows that $E(\lambda)$ has the transversal property if

$$\lambda = \lim \lambda_j \qquad (j = 1,2, \cdots ; 0 \leqq \lambda_j \leqq 1)$$

and each family $E(\lambda_j)(j = 1,2, \cdots)$ has the transversal property. Since the transversal property is possessed by $E(1) = E$, we conclude that there is a smallest value $\lambda = \rho \geqq 0$ for which $E(\lambda)$ has the transversal property. If $\rho = 0$, $E(\rho)$ consists of a finite number of points, and because of the transversal property these all lie on a line. Obviously this line intersects each of the ovals, namely at the chosen center of contraction. If $\rho > 0$, there are three ovals $A_i, A_k, A_l \in E(i < k < l)$ and a line L such that L intersects the ovals $A'_i = \rho A_i, A'_k = \rho A_k, A'_l = \rho A_l$ in three points P_i, P_k, P_l in the given order but the sets $\lambda A_i, \lambda A_k, \lambda A_l$ lack the transversal property for each $\lambda < \rho$. If U and V denote the two closed half planes bounded by L, it can be verified that only one of the ovals A'_i and A'_k can include interior points Q_i or Q_k of U. In the contrary case there would be a line H through P_l that meets the segments P_iQ_i and P_kQ_k of U at inner points; further, there would be a $\lambda < \rho$ such that either H or a line parallel to H intersects the sets $\lambda A_i, \lambda A_k$, and λA_l in the given order; but this is impossible. For the same reasons, at most one of the ovals A'_k and A'_l can include inner points of U, and of course the same arguments apply also to V. In other words, A'_i and A' lie entirely in one of the half planes, say in U, while A'_k lies entirely in the other half plane V. Traversing a transversal of two of the ovals in the given order, we therefore conclude that (a) on each transversal of A'_i and A'_k, the points before A'_i lie in U and those after A'_k lie in V; (b) on each transversal of A'_k and A'_l the points before A'_k lie in V and those after A'_l lie in U; (c) on each transversal of A'_i and A'_l, the points between A'_i and A'_l lie in U. Now let us consider an arbitrary oval $A'_j \in E(\rho)(j \neq i,k,l)$. If $j < i$ then by (a) A'_j includes points of U, for (A'_j, A'_i, A'_k) has the transversal property; by (b) A'_j includes points of V, for (A'_j, A'_k, A'_l) also has the transversal property. When $i < j < k$ or $k < j < l < j$ it follows similarly from (a), (b), and (c) that A'_j has points in common with both U and V and hence with the line L. Thus L intersects all the ovals of $E(\rho)$ and a fortiori all the ovals of E. When E is countably infinite, it follows from a supplementary convergence argument that all the ovals of E can be intersected by some line if this is true for each finite subfamily of E.

90. The numbering of the ovals provides the family of ovals with an order relation. If i,k, and l are three natural numbers with $i < k < l$, we have $A_i \subset C_i, A_k \subset D_i, A_k \subset C_k, A_l \subset D_k, C_i \subset C_k$, and $D_i \supset D_k$. If L and L' are intersecting lines that respectively separate C_i from D_i and C_k from D_k, then A_i and A_l lie in two opposite angles formed by L and L' while A_k lies in one of the other two angles. If a line H intersects the three ovals, then the intersection $H \cap A_k$ lies between the segments $H \cap A_i$ and $H \cap A_l$. The same considerations apply when L and L' are parallel; then A_k lies in a strip while A_i and A_l are in the two adjacent half planes. In each case the line H, suitably oriented, intersects the ovals in the prescribed order, and thus the result is a corollary of Proposition **89**.

91. The hypotheses imply that the family consists of countably many ovals $A_i (i = 1,2, \cdots)$ and that there is a point Z that does not belong to any of them. By means of the rays that emanate from Z and intersect a given A_i, the oval A_i is projected onto an arc H_i of the unit circle C centered at Z; H_i represents the angle that A_i subtends at Z. Since the ovals are all assumed to be proper and also to be pairwise congruent and disjoint, only finitely many of the countably many ovals can lie inside a fixed circle centered at Z. Thus the length of the arc H_i must tend to zero as i tends to infinity. By the Bolzano-Weierstrass theorem the unit circle C includes a point P, which we shall call the pole, such that the circular arcs $\tilde{H}_i$ of a suitably chosen subsequence of the H_i converge to the pole P as $i \to \infty$; thus for each $\epsilon > 0$ there is an n such that for $i > n$ the arcs $\tilde{H}_i$ lie entirely within an ϵ neighborhood of the pole P. Now let us choose two arbitrary ovals A_x and A_y from the original family of ovals. If $\tilde{A}_i (i = 1,2, \cdots)$ is the sequence of ovals corresponding to the sequence $\tilde{H}_i$ of arcs, then there exists m such that for all $i > m$, $\tilde{A}_i$ is different from both A_x and A_y. By hypothesis there is a line L_i that intersects the three ovals A_x, A_y, and $\tilde{A}_i$. We orient this line so that it first meets A_x and later $\tilde{A}_i$; this can be done in a unique way. Then we introduce the unit vector that points in the direction of the oriented line L_i and goes from Z to the point Q_i. The situation is such that the points Q_i of the circle C converge to the pole P as $i \to \infty$. Since the sets A_x and A_y are closed and bounded, there is a line in the direction represented by P that intersects both A_x and A_y. Thus the family of ovals has the property that each one of two of its members admits a transversal parallel to a fixed direction. If we project the ovals orthogonally onto a line that is perpendicular to the given direction, the resulting segments must intersect pairwise. Then by Proposition **16**, the simplest special case of Helly's theorem, all the segments must have a common point, and the projecting line corresponding to this point intersects all the ovals $A_i (i = 1,2, \cdots)$. This completes the proof.

92. The set of all points at which a circle subtends an angle of at least $\pi/3$ is a concentric disk having twice the radius. Thus the hypothesis that the disks with doubled radius are disjoint implies that at no point of the plane does more than one of the disks subtend an angle $\geqq \pi/3$. Consequently the result is a corollary of Proposition **28**.

93. Let us assume that the hypotheses are satisfied. From this it follows at once that the family of ovals is countable and that each bounded part of the plane contains only finitely many of them. Let us assume further, contrary to the claimed result, that the family of ovals is not extremely sparsely distributed. Then there exist positive numbers c and R such that $N(R) \geqq cR$ for all $R > R_0$. In each of the ovals $A_i (i = 1,2, \cdots)$ of the family there can be inscribed a circle of fixed positive radius ρ, and the ovals may be numbered so that the distances d_i of the centers of these inscribed circles from the reference point Z form a monotonically nondecreasing sequence. Let H_i be the circular arc of length α_i that is cut from the unit circle C about Z by the angle that A_i subtends at Z. Then $\alpha_i \geqq 2\rho/d_i$. Now it is easily established that the series $\Sigma_i^\infty (1/d_i)$ diverges. To see this, choose first an arbitrary $d > 0$ and let m denote the number of d_i's for which $d_i \leqq d$. Then choose an $R > R_0$ such that $2m/R < c$, where c and R_0 are the constants introduced earlier. Let M be the number of d_i's for which $d_i \leqq R$. Then always $M \geqq N(R) \geqq cR$, by definition of the function $N(R)$. Now we consider the partial sum

$$\left(\frac{1}{d_{m+1}}\right) + \cdots + \left(\frac{1}{d_M}\right) = s$$

and find easily that

$$s \geqq \frac{M - m}{R} \geqq \frac{c}{2}.$$

Since m is arbitrarily large for a sufficiently large value of d, our series cannot converge.

If n is the natural number appearing in the hypothesis and p is chosen so large that

$$2\rho\left(\frac{1}{d_i} + \cdots + \frac{1}{d_p}\right) \geqq 2n\pi,$$

then there must be a point P of the circle C that belongs to at least n different arcs $H_i (i = 1, \cdots, p)$. The line through Z and P meets at least n of the ovals and this contradicts the hypothesis.

94. Let us consider a line L in the given direction. If it or some parallel line intersects infinitely many ovals of the family, the desired conclusion holds. Thus we assume henceforth that each line parallel to L intersects only finitely many, but at least one, of the ovals. Among the ovals intersected by L, let A be one whose supporting lines L_1 and L_2 parallel to L have maximum distance. Let $H_i (i = 1,2)$ be a half plane that contains A and is bounded by L_i. Then L_i must intersect at least one oval A_i of the family that is not entirely contained in H_i; otherwise there would be a nearby parallel of L_i that meets infinite ovals of the family or none at all, both of which are excluded. Now we show that among A, A_1, A_2 and the ovals of the family that intersect H_1, there must always be three ovals that admit a common transversal. By construction, no two of the sets A, A_1, A_2 are identical. If this triple admits a common transversal, the claim is established. If not, the three sets are clearly pairwise disjoint and the ovals A and A_1 admit two common inner supporting lines T and T'. The point Q common to T and T' lies in H_1. If, among the infinitely many ovals of the family that lie in H_1, some one other than A includes a point P of the angle W that contains A and is determined by T and T', then the line PQ intersects that oval in addition to A

and A_1, and a triple of the desired sort is obtained. In the other case, the oval A_2, in particular, does not lie in W. Then if Q' is a point of A that is in the supporting line L_2, Q' lies in W. Let W' be the angle that is filled out by the rays that emanate from Q' and intersect A_2. If, among the ovals of the family that lie in H_1, at least one other than A and A_2 includes a point P' of W', then the line $P'Q'$ meets that oval in addition to A and A_2, whence a triple of the desired sort is obtained. There remains only the case in which all of the infinitely many ovals of the family other than A and A_2 lie entirely in the difference set $H_1 - W - W'$. This set contains an angle V whose bounding rays are not parallel to L, which lies entirely in H_1, and which with only finitely many exceptions contains all of the ovals of the family that lie in H_1. An obvious calculation shows that in this case the subfamily consisting of those ovals of the given family that lie in H_1 is not extremely sparsely distributed, so the hypothesis of Proposition **93** cannot be satisfied. Thus in particular there is a line that intersects at least three ovals of this subfamily. Fixing such a triple, we then consider a line L' that is interior to H_1, hence parallel to L, and has the triple entirely to one side of it. Applying to L' the reasoning undertaken initially for L, we find another triple of the desired sort and can continue in this way ad infinitum. This completes the proof.

95. Proposition is a corollary of Proposition **94**.

96. We divide the pairs of ovals from the family into two classes according to whether the two ovals of the pair have an empty or nonempty intersection. By Proposition **68** the family of ovals contains an infinite subfamily such that all its pairs belong to the same class. If, contrary to our claim, there were no infinite subfamily consisting of pairwise disjoint ovals, then there would be an infinite subfamily whose ovals are pairwise intersecting. If we project the ovals of this subfamily orthogonally onto a line T, the resulting segments intersect pairwise and hence by Proposition **16** have a common point P. The line L that is perpendicular to T at P intersects all the ovals of the subfamily, contradicting the hypothesis.

97. We divide the pairs of rectangles into two classes according to whether the two rectangles of the pair have an empty or nonempty intersection. By Proposition **68** the family of rectangles has an infinite subfamily whose pairs all belong to the same class. By hypothesis this can only be the second class and then the desired conclusion is obtained with the aid of Proposition **15**.

98. We divide the triples of ovals into two classes according to whether the three ovals of the triples have an empty or nonempty intersection. By Proposition **68** the family of ovals has an infinite subfamily whose triples all belong to the same class. In view of the hypotheses this cannot be the first class, whence by Proposition **14** the ovals of the subfamily have a nonempty intersection.

Bibliography

(Note that the Bibliography is in two sections, each arranged alphabetically. Items [1] to [114] appeared in the original text and are mentioned here in Chapters 1 to 10. Items [115] to [202] have been added by the translator and are mentioned here in Chapter 11.

[1] ALTWEGG, M., "Ein Satz über Mengen von Punkten mit ganzzahliger Entfernung." *Elemente der Math.*, **7**, pp. 56–58, 1952.

[2] ANNING, N. H., and P. ERDÖS, "Integral distances." *Bull. Am. Math. Soc.*, **51**, pp. 598–600, 1945.

[3] BALASUBRAMANIAN, N., "A theorem on sets of points." *Proc. Nat. Inst. Sci. India*, **19**, p. 839, 1953.

[4] BANG, Th., "On covering by parallel-strips." *Mat. Tidsskr. B.*, **1950**, pp. 49–53, 1950.

[5] ———, "A solution of the plank problem." *Proc. Am. Math. Soc.*, **2**, pp. 990–993, 1951.

[6] BERNHEIM, B., and Th. MOTZKIN, "A criterion for divisibility of *n*-gons into *k*-gons." *Comment. Math. Helvetici*, **22**, pp. 93–102, 1949.

[7] BLASCHKE, W. *Kreis and Kugel.* Leipzig: Veit & Co., 1916, and New York: Chelsea Publishing Company, 1949; Berlin: De Gruyter, 2nd ed., 1956.

[8] BORSUK, K., "Drei Sätze über die *n*-dimensionale euklidische Sphäre." *Fundamenta Math.*, 20, pp. 177–190, 1933.

[9] DE BRUIJN, N. G., and P. ERDÖS, "On a combinatorial problem." *Indagationes Math.*, **10**, pp. 421–423, 1948.

[10] COXETER, H. S. M., "A problem of collinear points." *Am. Math. Monthly*, **55**, pp. 26–28, 1948.

[11] DANZER, L., "Über ein Problem aus der kombinatorischen Geometrie." *Arch. der Math.*, **8**, pp. 347–351, 1957.

[12] DELACHET, A. *La Géométrie Contemporaine.* Paris: Presses Universitaire de France, 1950.

[13] DIRAC, G. A., "Collinearity properties of sets of points." *Quart. J. Math. Oxford*, Ser. (2), **2**, pp. 221–227, 1951.

[14] EGGLESTON, H. G., "Covering a three-dimensional set with sets of smaller diameter." *J. London Math. Soc.*, **30**, pp. 11–24, 1955.

[15] ———. *Problems in Euclidean Space: Application of Convexity.* London: Pergamon Press, 1957.

[16] ERDÖS, P., "Problem No. 4065." *Am. Math. Monthly*, **51**, pp. 169–171, 1944.

[17] ———, "Integral distances." *Bull. Am. Math. Soc.*, **51**, p. 996, 1945.

[18] ———, "On sets of distances of *n* points." *Am. Math. Monthly*, **53**, pp. 248–250, 1946.

[19] ———, "Some remarks on the theory of graphs." *Bull. Am. Math. Soc.*, **53**, pp. 292–294, 1947.

[20] ———, "Aufgabe 250." *Elemente der Math.*, **10**, p. 114, 1955. (Solutions by H. Debrunner, also by A. Bager, **11**, p. 137, 1956.)

[21] ———, and G. SZEKERES, "A combinatorial problem in geometry." *Compositio Math.*, **2**, pp. 463–470, 1935.

[22] FEJES TÓTH, L. *Lagerungen in der Ebene, auf der Kugel und im Raum.* Berlin: Springer, 1953.

[23] FENCHEL, W., "On Th. Bang's solution of the plank problem." *Mat. Tidsskr. B.*, **1951,** pp. 49–51, 1951.

[24] GALE, D., "On inscribing *n*-dimensional sets in a regular *n*-simplex." *Proc. Am. Math. Soc.*, **4,** pp. 222–225, 1953.

[25] GOODMAN, A. W., and R. E. GOODMAN, "A circle covering theorem." *Am. Math. Monthly*, **52,** pp. 494–498, 1945.

[26] GRÜNBAUM, B., "On a theorem of L. A. Santaló." *Pacific J. Math.*, **5,** pp. 351–359, 1955.

[27] ———, "A simple proof of Borsuk's conjecture in three dimensions." *Proc. Cambridge Phil. Soc.*, **53,** pp. 776–778, 1957.

[28] ———, "On common transversals." *Arch. der Math.*, **9,** pp. 465–469, 1958.

[29] GUPTA, H., "Non-concyclic sets of points." *Proc. Nat. Inst. Sci. India*, **19,** pp. 315–316, 1953.

[30] GUSTIN, W., "On the interior of the convex hull of a Euclidean set." *Bull. Am. Math. Soc.*, **53,** pp. 299–301, 1947.

[31] HADWIGER, H., "Ein Überdeckungssatz für den euklidischen Raum. *Portugaliae Math.*, **4,** pp. 140–144, 1944.

[32] ———, Überdeckung des euklidischen Raumes durch kongruente Mengen." *Portugaliae Math.*, **4,** pp. 238–242, 1945.

[33] ———, "Über die rationalen Hauptwinkel der Goniometrie." *Elemente der Math.*, **1,** pp. 98–100, 1946.

[34] ———, Bemerkung zu einer Grössenrelation bei Punktmengen." *Portugaliae Math.*, **6,** pp. 45–48, 1947.

[35] ———, "Nonseparable convex systems." *Am. Math. Monthly*, **54,** pp. 583–585, 1947.

[36] ———, "Eulers Charakteristik und kombinatorische Geometrie." *J. reine angew. Math.*, **194,** pp. 101–110, 1955.

[37] ———, "Über einen Satz Hellyscher Art." *Arch. der Math.*, **7,** pp. 377–379, 1956.

[38] ———, *Wiskundige Opgaven* (Amsterdam, 1956), **20,** p. 3, 1957.

[39] ———, "Über Eibereiche mit gemeinsamer Treffgeraden." *Portugaliae Math.*, **16,** pp. 23–29, 1957.

[40] ———, and H. DEBRUNNER, "Über eine Variante zum Hellyschen Satz." *Arch. der Math.*, **8,** pp. 309–313, 1957.

[41] HANNER, O., and H. RÅDSTRÖM, "A generalization of a theorem of Fenchel." *Proc. Am. Math. Soc.*, **2,** pp. 589–593, 1951.

[42] HELLY, E., "Über Mengen konvexer Körper mit gemeinschaftlichen Punkten." *Jber. Deutsch. Math. Verein.*, **32,** pp. 175–176, 1923.

[43] HEPPES, A., "On the partitioning of three-dimensional point-sets into sets of smaller diameter" (Hungarian). *Magyar Tud. Akad. Mat. Fiz. Oszt. Közl*, **7,** pp. 413–416, 1957.

[44] ———, "Letter of September 9, 1958."

[45] ———, and P. RÉVÉSZ, "Zum Borsukschen Zerteilungsproblem." *Acta Math. Acad. Sci. Hungaricae*, **7,** pp. 159–162, 1956.

[46] HOPF, H., "Über die Sehnen ebener Kontinuen und die Schleifen geschlossener Wege." *Comment. Math. Helvetici*, **9,** pp. 303–319, 1936–1937.

[47] ———, "Über Zusammenhänge zwischen Topologie und Metrik im Rahmen der elementaren Geometrie." *Math. Phys. Semesterber.*, **3,** pp. 16–29, 1953.

[48] ———, and E. PANNWITZ, "Aufgabe Nr. 167." *Jber. Deutsch. Math. Verein,* **43,** p. *114*, 1934; **45,** p. *33*, 1935.

[49] HORN, A., "Some generalizations of Helly's theorem on convex sets." *Bull. Am. Math. Soc.*, **55,** pp. 923–929, 1949.

[50] ———, and F. A. VALENTINE, "Some properties of *L*-sets in the plane." *Duke Math. J.*, **16,** pp. 131–140, 1949.

[51] JUNG, H. W. E., "Über die kleinste Kugel, die eine räumliche Figur einschliesst." *J. reine angew. Math.*, **123,** pp. 241–257, 1901.

[52] KARLIN, S., and L. S. SHAPLEY, "Some applications of a theorem on convex functions." *Ann. Math.* (2), **52,** pp. 148–153, 1950.

[53] KELLY, L. M., "Covering problems." *Nat. Math. Mag.*, **19,** pp. 123–130, 1944.

[54] KHINTCHINE, A. *Three Pearls of Number Theory* (Russian). OGIZ, Moscow-Leningrad 1947. Translated into German by W. von Klemm, Berlin: Akademie-Verlag, 1951; into English by F. Bagemihl, H. Komm, and W. Seidel, Rochester, New York: Graylock Press, 1952.

[55] KIRCHBERGER, P., "Über Tschebyschefsche Annäherungsmethoden." *Math. Ann.*, **57,** pp. 509–540, 1903.

[56] KLEE, V. L., JR., "On certain intersection properties of convex sets." *Can. J. Math.*, **3,** pp. 272–275, 1951.

[57] ———, Letter to H. Hadwiger on February 20, 1953.

[58] ———, "The critical set of a convex body." *Am. J. Math.* **75,** pp. 178–188, 1953.

[59] ———, Letter to P. Vincensini on September 27, 1954.

[60] ———, "Common secants for plane convex sets." *Proc. Am. Math. Soc.*, **5,** pp. 639–641, 1954.

[61] KNESER, H., and W. SÜSS, "Aufgabe Nr. 299." *Jber. Deutsch. Math. Ver.*, **51,** p. *3*, 1941. (Solution by E. Hopf, **53,** p. *40*, 1943.)

[62] KNESER, M., Aufgabe Nr. 360. *Jber. Deutsch. Math. Verein.*, **58,** p. *27*, 1956. (Solution by M. Kneser **59,** p. *57*, 1956.)

[63] KÖNIG, D., "Über Graphen und ihre Anwendung auf Determinantentheorie und Mengenlehre." *Math. Ann.*, **77,** pp. 453–465, 1916.

[64] ———, "Über konvexe Körper." *Math. Z.*, **14,** pp. 208–210, 1922.

[65] KRASNOSSELSKY, M., "Sur un critère pour qu'un domaine soit étoilé." (Russian, with French summary.) *Mat. Sbornik, n. Ser.*, **19** (61), pp. 309–310, 1946.

[66] KUIPER, N. H., "On convex sets and lines in the plane." *Proc. Kon. Nederl. Akad. Wet.*, A **60,** pp. 272–283, 1957.

[67] LENZ, H., "Zur Zerlegung von Punktmengen in solche kleineren Durchmessers." *Arch. der Math.*, **6,** pp. 413–416, 1955.

[68] ———, Über die Bedeckung ebener Punktmengen durch solche kleineren Durchmessers." *Arch. der Math.*, **7,** pp. 34–40, 1956.

[69] LEVI, F. W., "On Helly's theorem and the axioms of convexity." *J. Indian Math. Soc.*, **15,** pp. 65–76, 1951.

[70] ———, "Eine Ergänzung zum Hellyschen Satze." *Arch. der Math.*, **4,** pp. 222–224, 1953.

[71] ———, "Überdeckung eines Eibereiches durch Parallelverschiebung seines offenen Kerns." *Arch. der Math.*, **6,** pp. 369–370, 1955.

[72] LÉVY, P., "Sur une généralisation du théorème de Rolle." *C. R. Acad. Sci. Paris*, **198**, pp. 424–425, 1934.

[73] MAAK, W., "Ein Problem der Kombinatorik in seiner Formulierung von H. Weyl." *Math. Phys. Semesterber*, **2**, pp. 251–256, 1952.

[74] MOLNÁR, J., "Über Sternpolygone." *Publ. Math. Debrecen*, **5**, pp. 241–245, 1958.

[75] MOSER, L., "On the different distances determined by *n* points." *Am. Math. Monthly*, **59**, pp. 85–91, 1952.

[76] MOTZKIN, Th., "The lines and planes connecting the points of a finite set." *Trans. Am. Math. Soc.*, **70**, pp. 451–464, 1951.

[77] MÜLLER, A., "Auf einem Kreis liegende Punktmengen ganzzahliger Entfernungen." *Elemente der Math.*, **8**, pp. 37–38, 1953.

[78] NAGY, B. SZ.-. "Ein Satz über Parallelverschiebungen konvexer Körper." *Acta Sci. Math.*, **15**, pp. 169–177, 1954.

[79] NEUMANN, J. VON, "Zur allgemeinen Theorie des Masses." *Fundamenta Math.*, **13**, pp. 73–116, 1929.

[80] OHMANN, D., "Kurzer Beweis einer Abschatzung fur die Breite bei Überdeckung durch konvexe Körper." *Arch. der Math.*, **8**, pp. 150–152, 1957.

[81] PÁL, J., "Über ein elementares Variationsproblem." *Math.-fys. Medd., Danske Vid. Selsk.*, **3**, 1920.

[82] PÓLYA, G., and G. SZEGÖ. *Aufgaben und Lehrsätze aus der Analysis*, **1, 2.** Berlin: Springer 1925, and New York: Dover, 1945.

[83] RADEMACHER, H., and I. J. SCHOENBERG, "Helly's theorems on convex domains and Tchebycheff's approximation problem." *Can. J. Math.*, **2**, pp. 245–256, 1950.

[84] ———, and O. TOEPLITZ. *Von Zahlen und Figuren.* Berlin: Springer, 1930 (2d ed., 1933), English translation by H. S. Zuckerman (with two additional chapters) as *The Enjoyment of Mathematics*, Princeton: Princeton University Press, 1957.

[85] RADO, R., "Verallgemeinerung eines Satzes von van der Waerden mit Anwendungen auf ein Problem der Zahlentheorie." *Sitzungsber. Preuss. Akad. Wiss. Berlin* (H. 16/17), pp. 589–596, 1933.

[86] ———, "Theorems on the intersection of convex sets of points." *J. London Math. Soc.*, **27**, pp. 320–328, 1952.

[87] ———, "A theorem on sequences of convex sets." *Quart. J. Oxford, Ser.* (2), **3**, pp. 183–186, 1952.

[88] RADON, J., "Mengen konvexer Körper, die einen gemeinsamen Punkt enthalten." *Math. Ann.*, **83**, pp. 113–115, 1921.

[89] RAMSEY, F. P., "On a problem of formal logic." *Proc. London Math. Soc.* (2), **30**, pp. 264–286, 1929.

[90] ROBINSON, C. V., "Spherical theorems of Helly type and congruence indices of spherical caps." *Am. J. Math.*, **64**, pp. 260–272, 1942.

[91] SANTALÓ, L. A., "Un teorema sobre conjuntos de paralelepipedos de aristas paralelas." *Publ. Inst. Mat. Univ. Nac. Litoral*, **2**, pp. 49–60, 1940; **3**, pp. 202–210, 1942.

[92] ———, "Sobre pares de figuras convexas." *Gaz. Mat. Lisboa*, **12**, pp. 7–10, 1951; **14**, p. 6, 1953.

[93] SCHERRER, W., "Die Einlagerung eines regularen Vielecks in ein Gitter." *Elemente der Math.*, **1,** pp. 97–98, 1946.

[94] SIERPINSKI, W., *On the congruence of sets and their equivalence by finite decomposition.*" *Lucknow Univ. Studies*, **20,** 1954.

[95] SKOLEM, Th., "Ein kombinatorischer Satz mit Anwendung auf ein logisches Entscheidungsproblem." *Fundamenta Math.*, **20,** pp. 254–261, 1933.

[96] SPERNER, E., "Note zu der Arbeit von Herrn B. L. van der Waerden: Ein Satz über Klasseneinteilungen von endlichen Mengen." *Abh. Math. Sem. Univ. Hamburg*, **5,** p. 232, 1927.

[97] ———, "Neuer Beweis für die Invarianz der Dimensionszahl und des Gebietes." *Abh. Math. Sem. Univ. Hamburg*, **6,** pp. 265–272, 1928.

[98] STEIGER, F., "Zu einer Frage über Mengen von Punkten mit ganzzahliger Entfernung." *Elemente der Math.*, **8,** pp. 66–67, 1953.

[99] STEINITZ, E., "Bedingt konvergente Reihen und konvexe Systeme." *J. reine angew. Math.*, **143,** pp. 128–175, 1913; **144,** pp. 1–40, 1914; **146,** pp. 1–52, 1916.

[100] STRASZEWICZ, S., "Un théorème sur la largeur des ensembles convexes." *Ann. Soc. Polonaise Math.*, **21,** pp. 90–93, 1948.

[101] STRAUS, E. G., "On a problem of W. Sierpinski on the congruence of sets." *Fundamenta Math.* **44,** pp. 75–81, 1957.

[102] SYLVESTER, J. J., "Question No. 11851." *Educational Times*, **59,** p. 98, 1893.

[103] TREVISAN, G., "Una condizione di allineamento per gli insiemi infiniti di punti del piano euclideo." *Rend. Seminar. Mat. Univ. Padova*, **18,** 258–261, 1949.

[104] TROST, E., "Bemerkung zu einem Satz über Mengen von Punkten mit ganzzahliger Entfernung." *Elemente der Math.*, **6,** pp. 59–60, 1951.

[105] VALENTINE, F. A., "A three point convexity property." *Pacific J. Math.* **7,** pp. 1227–1235, 1957.

[106] VAN DER WAERDEN, B. L., "Ein Satz über Klasseneinteilungen von endlichen Mengen." *Abh. Math. Sem. Univ. Hamburg*, **5,** pp. 185–188, 1927.

[107] ———, "Beweis einer Baudetschen Vermutung." *Nieuwe Arch. Wiskunde* (2), **15,** 212–216, 1927.

[108] ———, "Aufgabe Nr. 51." *Elemente Math.*, **4,** p. 18, 1949. (Solution by W. Gysin, also by D. Puppe and J. M. Ebersold, **4,** p. 140, 1949.)

[109] ———, "Einfall und Überlegung in der Mathematik (dritte Mitteilung). Der Beweis der Vermutung von Baudet." *Elemente Math.*, **9,** pp. 49–56, 1954.

[110] VINCENSINI, P., "Sur une extension d'un théorème de M. J. Radon sur les ensembles de corps convexes." *Bull. Soc. Math. France*, **67,** pp. 115–119, 1939.

[111] ———, "Les ensembles d'arcs d'un même cercle dans leurs relations avec les ensembles de corps connexes du plan euclidien." *Atti IV. Congr. Un. Mat. Ital.*, **2,** pp. 456–464, 1953.

[112] ———, "Sur certains ensembles d'arcs de cercle ou de calottes sphériques." *Bull. Sci. Math.*, (2), **77,** pp. 120–128, 1953.

[113] WITT, E., "Ein kombinatorischer Satz der Elementargeometrie." *Math. Nachr.*, **6,** pp. 261–262, 1951.

[114] YAGLOM, I. M., and W. G. BOLTYANSKIĬ. *Convex Figures* (Russian). Moscow-Leningrad: Gosudarstv. Izdat. Tehn.-Teor. Lit., 1951. Translated into German by H. Grell, K. Maruhn, and W. Rinow, Berlin: VEB Deut-

scher Verlag der Wissenschaften 1956; into English by P. J. Kelly and L. F. Walton, New York: Holt, Rinehart, and Winston, Inc., 1961.

[115] BANACH, S., "Sur le problème de mesure." *Fund. Math.*, **4,** pp. 7–33, 1923.

[116] ———, and A. TARSKI, "Sur la decomposition des ensembles de points en parties respectivement congruents." *Fund. Math.*, **6,** pp. 244–277, 1924.

[117] BERGE, C. *Théorie des Graphes et Ses Applications.* Paris: Dunod, 1958. (English translation by A. Doigs. London: Methuen & Co., Ltd., and New York: John Wiley and Sons, Inc., 1962

[118] ———, and A. GHOUILA-HOURI. *Programmes, jeux et reseaux de transport.* Paris: Dunod, 1962.

[119] BIRCH, B. J., "On $3N$ points in a plane." *Proc. Cambridge Phil. Soc.*, **55,** pp. 289–293, 1959.

[120] BOLTYANSKIĬ, V. G. *Equivalent and Equidecomposable Figures.* Boston: Heath, 1963. (Translated by A. K. Henn and C. E. Watts from the first Russian edition, Moscow, 1956.)

[121] BONNESEN, T., and W. FENCHEL. *Theorie der Konvexen Körper.* Berlin: Springer, 1934. (Reprint, New York: Chelsea Publishing Company, 1948.)

[122] BONNICE, W., and V. KLEE, "The generation of convex hulls." *Math. Ann.*, **152,** to appear, 1963.

[123] BOURGIN, D. G., "Some mapping theorems." *Rend. Mat. e Appl.* (5), **15,** pp. 177–189, 1956.

[124] ———, "Deformation and mapping theorems." *Fund. Math.*, **46,** pp. 285–303, 1959.

[125] BRUIJN, N. G. DE, and P. ERDÖS, "A colour problem for infinite graphs and a problem in the theory of relations." *Indag. Math.*, **13,** pp. 369–373, 1951.

[126] CARATHÉODORY, C., "Uber den Variabilitätsbereich der Koeffizienten von Potenzreihen, die gegebene Werte nicht annehmen." *Math. Ann.*, **64,** 95–115, 1907.

[127] CHRESTENSON, H. E., and M. S. KLAMKIN, "Polygon imbedded in a lattice." *Am. Math. Monthly*, **70,** pp. 447–448, 1963.

[128] COXETER, H. S. M. *Regular Polytopes*, 2nd ed. New York: Macmillan, 1963.

[129] DANZER, L., "Über Durchschnittseigenschaften n-dimensionaler Kugelfamilien." *J. Reine Angew. Math.*, **208,** pp. 181–203, 1961.

[130] ———, B. GRÜNBAUM, and V. KLEE., "Helly's theorem and its relatives." Proceedings of Symposia in Pure Mathematics, Am. Math. Soc., **7,** *Convexity*, pp. 100–181, 1963.

[131] DEKKER, T. J. *Paradoxical Decompositions of Sets and Spaces.* Thesis, University of Amsterdam, 1958.

[132] EGGLESTON, H. G. *Convexity*, London: Cambridge, 1958.

[133] ERDÖS, P., "On some geometrical problems." *Mathematikai Lapok,* **8,** pp. 86–92, 1957.

[134] ———, "On sets of distances of n points in Euclidean space." *Magyar Tud. Akad. Mat. Kutató Int. Közl*, **5,** pp. 165–169, 1960.

[135] ———, and C. A. ROGERS, "Covering space with convex bodies." *Acta Arithm.*, **7,** pp. 281–285, 1962.

[136] ———, and G. SZEKERES, "On some extremum problems in elementary geometry." *Ann. Univ. Sci. Budapest Eötvös. Sect. Math.*, **3–4,** pp. 53–62, 1960/61.

[137] FLOYD, E. E., "Real-valued mappings of spheres." *Proc. Am. Math. Soc.*, **6,** pp. 957–959, 1955.

[138] FORD, L. R., JR., and D. R. FULKERSON. *Flows in Networks.* Princeton, N. J.: Princeton University Press, 1962.

[139] FRASNAY, C., "Sur les fonctions d'entiers se rapportant au théorème de Ramsay." *C. R. Acad. Sci. Paris*, **256,** pp. 2507–2510, 1963.

[140] FULLERTON, R. E., "Integral distances in Banach spaces." *Bull. Am. Math. Soc.*, **55,** pp. 901–905, 1949.

[141] GERICKE, H., "Über die grösste Kugel in einer konvexe Punktmenge." *Math. Z.*, **40,** pp. 317–320, 1936.

[142] GOHBERG, J. C., and A. S. MARKUS, "A problem on covering a convex figure by similar ones." (Russian). *Izv. Mold. Fil. Akad. Nauk. SSSR* **10** (76), pp. 87–95, 1960.

[143] GREENWOOD, R. E., and A. M. GLEASON, "Combinatorial relations and chromatic graphs." *Can. J. Math.*, **7,** pp. 1–7, 1955.

[144] GROEMER, H., "Abschätzungen für die Anzahl der konvexen Körper die einen konvexen Körper berühren." *Monatsh. Math.*, **65,** pp. 74–81, 1961.

[145] GRÜNBAUM, B., "A proof of Vázsonyi's conjecture." *Bull. Research Council Israel.* Sect. A., **6,** pp. 77–78, 1956.

[146] ———, "Borsuk's partition conjecture in Minkowski planes." *Bull. Research Council Israel.* Sect. F., **7,** pp. 25–30, 1957.

[147] ———, "On intersections of similar sets." *Portugal. Math.* **18,** pp. 155–164, 1959.

[148] ———, "Common transversals for families of sets." *J. London Math. Soc.*, **35,** 408–416, 1960.

[149] ———, "On a conjecture of Hadwiger." *Pacific J. Math.*, **11,** pp. 215–219, 1961.

[150] ———, "The dimension of intersections of convex sets." *Pacific J. Math.*, **12,** pp. 197–202, 1962.

[151] ———, "Borsuk's problem and related questions." Am. Math. Soc. Proceedings of Symposia in Pure Mathematics, **7,** *Convexity*, pp. 271–284, 1963.

[152] HADWIGER, H., "Von der Zerlegung der Kugel in kleinere Teile." *Gaz. Math. Lisboa*, **15,** pp. 1–3, 1954.

[153] ———. *Vorlesungen über Inhalt Oberfläche und Isoperimetrie.* Berlin: Springer, 1957.

[154] ———, "Ungelöste Probleme, No. 24." *Elemente der Math.*, **13,** p. 85, 1958.

[155] ———, "Elementare Begründung ausgewählter stetigkeitsgeometrischer Sätze für Kreis und Kugelfläche." *Elemente der Math.*, **14,** pp. 49–60, 1959.

[156] ———, "Elementare Kombinatorik und Topologie." *Elemente der Math.*, **15,** 49–60, 1960.

[157] ———, "Ein Satz über stetige Funktionen auf der Kugelfläche." *Arch. Math.*, **11,** pp. 65–68, 1960.

[158] ———, "Ungelöste Probleme, No. 40." *Elemente der Math.*, **16,** pp. 103–104, 1961.

[159] HALBERG, C. J. A., JR., E. LEVIN, and E. G. STRAUS, "On contiguous congruent sets in Euclidean space." *Proc. Am. Math. Soc.*, **10,** pp. 335–344, 1959.

[160] HANNER, O., "Intersections of translates of convex bodies." *Math. Scand.* **4**, pp. 65–87, 1956.

[161] HEPPES, A., "Beweis einer Vermutung von A. Vázsonyi." *Acta. Math. Acad. Sci. Hungar.*, **7**, pp. 463–466, 1957.

[162] HERZOG, F., and L. M. KELLY, "A generalization of a theorem of Sylvester." *Proc. Am. Math. Soc.*, **11**, pp. 327–331, 1960.

[163] HOPF, H., "Verallgemeinerung bekannter Abbildungs und Überdeckungssätze." *Portugal. Math.*, **4**, pp. 129–139, 1944.

[164] KAKUTANI, S., "A proof that there exists a circumscribing cube around any closed bounded convex set inR^3." *Ann. of Math.*, **43**, pp. 739–741, 1942.

[165] KELLY, L. M., and W. O. J. MOSER, "On the number of ordinary lines determined by *n* points." *Canad. J. Math.*, **10**, pp. 210–219, 1958.

[166] KLEE, V., "Convex sets in linear spaces." *Duke Math. J.*, **18**, pp. 443–466, 1953.

[167] ———, "Infinite-dimensional intersection theorems." Proceedings of Symposia in Pure Mathematics, Am. Math. Soc., **7**, *Convexity*, pp. 349–360, 1963.

[168] KNASTER, B., "Problem No. 4." *Colloq. Math.*, **1**, pp. 30–31, 1947.

[169] KÖNIG, D. *Theorie der endlichen und unendlichen Graphen.* Akad Verl. M.B.H., Leipziz, 1936 (Reprint New York: Chelsea Publishing Company, 1950.)

[170] KUHN, H. W., "Some combinatorial lemmas in topology." *IBM J. Res. Develop.*, **4**, pp. 508–524, 1960.

[171] LEKKERKERKER, C. G., and J. C. BOLAND, "Representation of a finite graph by a set of intervals on the real line." *Fund. Math.*, **51**, pp. 45–64, 1962.

[172] LIVESAY, G. R., "On a theorem of F. J. Dyson." *Ann. of Math.*, **59**, pp. 227–229, 1954.

[173] MESCHKOWSKI, H., *Ungelöste und unlösbare Probleme der Geometrie.* Brunswick, Germany: Vieweg-Verlag, 1960.

[174] MIRA FERNANDES, A. DE., "Funzioni continue sopra una superficie sferica." *Portugal. Math.*, **5**, pp. 132–134, 1946.

[175] MORSE, A. P., "Squares are normal." *Fund. Math.*, **36**, pp. 35–39, 1949.

[176] NEWMAN, D. J., and R. BREUSCH, "Covering a rectangle with two smaller similar rectangles." *Am. Math. Monthly*, **70**, pp. 450–451, 1963.

[177] ORE, O., *Theory of Graphs.* Am. Math. Soc. Colloq. Pub., **38**, 1962.

[178] RATTRAY, B. A., "An antipodal-point, orthogonal point theorem." *Ann. of Math.*, **60**, pp. 502–512.

[179] RIORDAN, J. *Introduction to Combinatorial Analysis.* New York: Wiley, 1958.

[180] ROBINSON, R. M., "On the decomposition of spheres." *Fund. Math.*, **34**, pp. 246–260, 1947.

[181] RYSER, H. J., *Combinatorial Mathematics.* Carus Mathematical Monographs, No. 14, Math. Assoc. Amer., 1963.

[182] SANTALÓ, L. A., "Convex figures on the *n*-dimensional spherical surface." *Ann. of Math.*, **47**, pp. 448–459, 1946.

[183] SANTIS, R. DE., "A generalization of Helly's theorem." *Proc. Am. Math. Soc.*, **8**, 336–340, 1957.

[184] SIERPINSKI, W., "Un théorème sur les continus." *Tôhoku Math. J.*, **13**, pp. 300–303, 1918.

[185] ———, W. "Sur quelques problèmes concernant la congruence des ensembles de points." *Elemente der Math.*, **5**, pp. 1–4, 1950.

[186] ———, "Sur les ensembles de points aux distances rationelles situés sur un cercle." *Elemente der Math.*, **14,** pp. 25–27, 1959.

[187] STEINHAGEN, P., "Über die grösste Kugel in einer konvexen Punktmenge." *Abh. Mat. Sem. Univ. Hamburg*, **1,** pp. 15–26, 1921.

[188] STRASCEWICZ, S., "Sur un problème géométrique de P. Erdös." *Bull. Acad. Polon. Sci. Cl. III*, **5,** pp. 39–40, 1957.

[189] TARSKI, A., "On the equivalence of polygons" (Polish). *Przeg. Matem.-Fiz.*, **2,** 1924.

[190] ———, "Problem No. 38." *Fund. Math.*, **7,** p. 381, 1925.

[191] TUCKER, A. W., "Some topological properties of disk and sphere." *Proc. First Canadian Math. Congress* (Montreal 1945), 285–309.

[192] ULAM, S. *A Collection of Mathematical Problems.* New York: Interscience, 1960.

[193] VALENTINE, F. A., "The intersection of two convex surfaces and property P_3." *Proc. Am. Math. Soc.*, **9,** pp. 47–54, 1958.

[194] ———, "Characterizations of convex sets by local support properties." *Proc. Am. Math. Soc.*, **11,** pp. 112–116, 1960.

[195] YAMABE, H., and Z. YUJOBÔ, "On the continuous functions defined on a sphere." *Osaka Math. J.*, **2,** pp. 19–22, 1950.

[196] YANG, C. T., "On theorems of Borsuk-Ulam, Kakutani-Yamabe-Yujobô, and Dyson. I and II." *Ann. of Math.* (2), **60,** pp. 262–282, 1954 and **62,** pp. 271–283, 1955.

[197] ———, "Continuous functions from spheres to Euclidean spaces." *Ann. of Math.* (2) **62,** pp. 284–292, 1955.

[198] ———, "On maps from spheres to Euclidean spaces." *Am. J. Math.*, **749,** pp. 725–732, 1957.

[199] Altman, E., "On a problem of Erdös." *Am. Math. Monthly*, **70,** pp. 148–157, 1963.

[200] DEBRUNNER, H., "Orthogonale Dreibeine in richtungsvollständigen, stetigen Geradenscharen des R^3." *Comm. Math. Helv.*, **37,** pp. 36–43, 1962.

[201] GERAGHTY, M. A., "Applications of Smith index to some covering and frame theorems." *Indag. Math.*, **23,** pp. 219–228, 1961.

[202] SONNEBORN, L. M., "Level sets on spheres." *Pacific J. Math.*, **13,** pp. 297–303, 1963.

Index

Index

The index involves only Part I of the text. For certain notions that are used with great frequency, the index lists only the page on which the term is first defined.

Mathematics-Bestsellers

HANDBOOK OF MATHEMATICAL FUNCTIONS: with Formulas, Graphs, and Mathematical Tables, Edited by Milton Abramowitz and Irene A. Stegun. A classic resource for working with special functions, standard trig, and exponential logarithmic definitions and extensions, it features 29 sets of tables, some to as high as 20 places. 1046pp. 8 x 10 1/2. 0-486-61272-4

ABSTRACT AND CONCRETE CATEGORIES: The Joy of Cats, Jiri Adamek, Horst Herrlich, and George E. Strecker. This up-to-date introductory treatment employs category theory to explore the theory of structures. Its unique approach stresses concrete categories and presents a systematic view of factorization structures. Numerous examples. 1990 edition, updated 2004. 528pp. 6 1/8 x 9 1/4. 0-486-46934-4

MATHEMATICS: Its Content, Methods and Meaning, A. D. Aleksandrov, A. N. Kolmogorov, and M. A. Lavrent'ev. Major survey offers comprehensive, coherent discussions of analytic geometry, algebra, differential equations, calculus of variations, functions of a complex variable, prime numbers, linear and non-Euclidean geometry, topology, functional analysis, more. 1963 edition. 1120pp. 5 3/8 x 8 1/2. 0-486-40916-3

INTRODUCTION TO VECTORS AND TENSORS: Second Edition--Two Volumes Bound as One, Ray M. Bowen and C.-C. Wang. Convenient single-volume compilation of two texts offers both introduction and in-depth survey. Geared toward engineering and science students rather than mathematicians, it focuses on physics and engineering applications. 1976 edition. 560pp. 6 1/2 x 9 1/4. 0-486-46914-X

AN INTRODUCTION TO ORTHOGONAL POLYNOMIALS, Theodore S. Chihara. Concise introduction covers general elementary theory, including the representation theorem and distribution functions, continued fractions and chain sequences, the recurrence formula, special functions, and some specific systems. 1978 edition. 272pp. 5 3/8 x 8 1/2. 0-486-47929-3

ADVANCED MATHEMATICS FOR ENGINEERS AND SCIENTISTS, Paul DuChateau. This primary text and supplemental reference focuses on linear algebra, calculus, and ordinary differential equations. Additional topics include partial differential equations and approximation methods. Includes solved problems. 1992 edition. 400pp. 7 1/2 x 9 1/4. 0-486-47930-7

PARTIAL DIFFERENTIAL EQUATIONS FOR SCIENTISTS AND ENGINEERS, Stanley J. Farlow. Practical text shows how to formulate and solve partial differential equations. Coverage of diffusion-type problems, hyperbolic-type problems, elliptic-type problems, numerical and approximate methods. Solution guide available upon request. 1982 edition. 414pp. 6 1/8 x 9 1/4. 0-486-67620-X

VARIATIONAL PRINCIPLES AND FREE-BOUNDARY PROBLEMS, Avner Friedman. Advanced graduate-level text examines variational methods in partial differential equations and illustrates their applications to free-boundary problems. Features detailed statements of standard theory of elliptic and parabolic operators. 1982 edition. 720pp. 6 1/8 x 9 1/4. 0-486-47853-X

LINEAR ANALYSIS AND REPRESENTATION THEORY, Steven A. Gaal. Unified treatment covers topics from the theory of operators and operator algebras on Hilbert spaces; integration and representation theory for topological groups; and the theory of Lie algebras, Lie groups, and transform groups. 1973 edition. 704pp. 6 1/8 x 9 1/4. 0-486-47851-3

Mathematics–Algebra and Calculus

VECTOR CALCULUS, Peter Baxandall and Hans Liebeck. This introductory text offers a rigorous, comprehensive treatment. Classical theorems of vector calculus are amply illustrated with figures, worked examples, physical applications, and exercises with hints and answers. 1986 edition. 560pp. 5 3/8 x 8 1/2. 0-486-46620-5

ADVANCED CALCULUS: An Introduction to Classical Analysis, Louis Brand. A course in analysis that focuses on the functions of a real variable, this text introduces the basic concepts in their simplest setting and illustrates its teachings with numerous examples, theorems, and proofs. 1955 edition. 592pp. 5 3/8 x 8 1/2. 0-486-44548-8

ADVANCED CALCULUS, Avner Friedman. Intended for students who have already completed a one-year course in elementary calculus, this two-part treatment advances from functions of one variable to those of several variables. Solutions. 1971 edition. 432pp. 5 3/8 x 8 1/2. 0-486-45795-8

METHODS OF MATHEMATICS APPLIED TO CALCULUS, PROBABILITY, AND STATISTICS, Richard W. Hamming. This 4-part treatment begins with algebra and analytic geometry and proceeds to an exploration of the calculus of algebraic functions and transcendental functions and applications. 1985 edition. Includes 310 figures and 18 tables. 880pp. 6 1/2 x 9 1/4. 0-486-43945-3

BASIC ALGEBRA I: Second Edition, Nathan Jacobson. A classic text and standard reference for a generation, this volume covers all undergraduate algebra topics, including groups, rings, modules, Galois theory, polynomials, linear algebra, and associative algebra. 1985 edition. 528pp. 6 1/8 x 9 1/4. 0-486-47189-6

BASIC ALGEBRA II: Second Edition, Nathan Jacobson. This classic text and standard reference comprises all subjects of a first-year graduate-level course, including in-depth coverage of groups and polynomials and extensive use of categories and functors. 1989 edition. 704pp. 6 1/8 x 9 1/4. 0-486-47187-X

CALCULUS: An Intuitive and Physical Approach (Second Edition), Morris Kline. Application-oriented introduction relates the subject as closely as possible to science with explorations of the derivative; differentiation and integration of the powers of x; theorems on differentiation, antidifferentiation; the chain rule; trigonometric functions; more. Examples. 1967 edition. 960pp. 6 1/2 x 9 1/4. 0-486-40453-6

ABSTRACT ALGEBRA AND SOLUTION BY RADICALS, John E. Maxfield and Margaret W. Maxfield. Accessible advanced undergraduate-level text starts with groups, rings, fields, and polynomials and advances to Galois theory, radicals and roots of unity, and solution by radicals. Numerous examples, illustrations, exercises, appendixes. 1971 edition. 224pp. 6 1/8 x 9 1/4. 0-486-47723-1

AN INTRODUCTION TO THE THEORY OF LINEAR SPACES, Georgi E. Shilov. Translated by Richard A. Silverman. Introductory treatment offers a clear exposition of algebra, geometry, and analysis as parts of an integrated whole rather than separate subjects. Numerous examples illustrate many different fields, and problems include hints or answers. 1961 edition. 320pp. 5 3/8 x 8 1/2. 0-486-63070-6

LINEAR ALGEBRA, Georgi E. Shilov. Covers determinants, linear spaces, systems of linear equations, linear functions of a vector argument, coordinate transformations, the canonical form of the matrix of a linear operator, bilinear and quadratic forms, and more. 387pp. 5 3/8 x 8 1/2. 0-486-63518-X

Mathematics-Geometry and Topology

PROBLEMS AND SOLUTIONS IN EUCLIDEAN GEOMETRY, M. N. Aref and William Wernick. Based on classical principles, this book is intended for a second course in Euclidean geometry and can be used as a refresher. More than 200 problems include hints and solutions. 1968 edition. 272pp. 5 3/8 x 8 1/2. 0-486-47720-7

TOPOLOGY OF 3-MANIFOLDS AND RELATED TOPICS, Edited by M. K. Fort, Jr. With a New Introduction by Daniel Silver. Summaries and full reports from a 1961 conference discuss decompositions and subsets of 3-space; n-manifolds; knot theory; the Poincaré conjecture; and periodic maps and isotopies. Familiarity with algebraic topology required. 1962 edition. 272pp. 6 1/8 x 9 1/4. 0-486-47753-3

POINT SET TOPOLOGY, Steven A. Gaal. Suitable for a complete course in topology, this text also functions as a self-contained treatment for independent study. Additional enrichment materials make it equally valuable as a reference. 1964 edition. 336pp. 5 3/8 x 8 1/2. 0-486-47222-1

INVITATION TO GEOMETRY, Z. A. Melzak. Intended for students of many different backgrounds with only a modest knowledge of mathematics, this text features self-contained chapters that can be adapted to several types of geometry courses. 1983 edition. 240pp. 5 3/8 x 8 1/2. 0-486-46626-4

TOPOLOGY AND GEOMETRY FOR PHYSICISTS, Charles Nash and Siddhartha Sen. Written by physicists for physics students, this text assumes no detailed background in topology or geometry. Topics include differential forms, homotopy, homology, cohomology, fiber bundles, connection and covariant derivatives, and Morse theory. 1983 edition. 320pp. 5 3/8 x 8 1/2. 0-486-47852-1

BEYOND GEOMETRY: Classic Papers from Riemann to Einstein, Edited with an Introduction and Notes by Peter Pesic. This is the only English-language collection of these 8 accessible essays. They trace seminal ideas about the foundations of geometry that led to Einstein's general theory of relativity. 224pp. 6 1/8 x 9 1/4. 0-486-45350-2

GEOMETRY FROM EUCLID TO KNOTS, Saul Stahl. This text provides a historical perspective on plane geometry and covers non-neutral Euclidean geometry, circles and regular polygons, projective geometry, symmetries, inversions, informal topology, and more. Includes 1,000 practice problems. Solutions available. 2003 edition. 480pp. 6 1/8 x 9 1/4. 0-486-47459-3

TOPOLOGICAL VECTOR SPACES, DISTRIBUTIONS AND KERNELS, François Trèves. Extending beyond the boundaries of Hilbert and Banach space theory, this text focuses on key aspects of functional analysis, particularly in regard to solving partial differential equations. 1967 edition. 592pp. 5 3/8 x 8 1/2. 0-486-45352-9

INTRODUCTION TO PROJECTIVE GEOMETRY, C. R. Wylie, Jr. This introductory volume offers strong reinforcement for its teachings, with detailed examples and numerous theorems, proofs, and exercises, plus complete answers to all odd-numbered end-of-chapter problems. 1970 edition. 576pp. 6 1/8 x 9 1/4. 0-486-46895-X

FOUNDATIONS OF GEOMETRY, C. R. Wylie, Jr. Geared toward students preparing to teach high school mathematics, this text explores the principles of Euclidean and non-Euclidean geometry and covers both generalities and specifics of the axiomatic method. 1964 edition. 352pp. 6 x 9. 0-486-47214-0